Decision Support for Water Framework Directive Implementation

Water Framework Directive Series

Decision Support for Water Framework Directive Implementation

Volume 3

Edited by
Peter A. Vanrolleghem

Publishing
London • New York

Published by
IWA Publishing, Alliance House, 12 Caxton Street, London SW1H 0QS, UK
Telephone: +44 (0)20 7654 5500; Fax: +44 (0)20 654 5555
Email: publications@iwap.co.uk; Web: www.iwapublishing.com

First published 2011

Cover design by www.designforpublishing.co.uk
Originated by The Manila Typesetting Company

British Library Cataloguing in Publication Data
A CIP catalogue record for this book is available from the British Library

ISBN 10: 1843393271
ISBN 13: 9781843393276

Contents

Edited by Peter Vanrolleghem. ISBN: 9781843393276. Published by IWA Publishing, London, UK.

Authors for this volume

Jaroslav Mysiak
Fondazione Eni Enrico Mattei (FEEM), and Euro-Mediterranean Center for Climate Change (CMCC)
Venice, Italy
E-mail: jaroslav.mysiak@feem.it

Guido M. Bazzani
National Research Council - Institute of Biometeorology (CNR IBIMET)
Bologna, Italy
E-mail: g.bazzani@ibimet.cnr.it

Marco Tamaro
Director of the Fondazione Benetton Studi Ricerche
Treviso, Italy
E-mail: marco.tamaro@fbsr.it

Edited by Peter Vanrolleghem. ISBN: 9781843393276. Published by IWA Publishing, London, UK.

Yaella Depietri
Institut de Ciència i Tecnologia Ambientals, Universitat Autònoma de Barcelona
Barcelona, Spain and
United Nations University, Institute for Environment and Human Security,
Bonn, Germany
E-mail: yaella.depietri@gmail.com

Carlo Giupponi
Università Ca' Foscari di Venezia, and Euro-Mediterranean Center for Climate
Change (CMCC)
Venezia, Italy
E-mail: cgiupponi@unive.it

Preface

Water resources planning and management and the development of appropriate policies require methodologies and tools that are able to support systematic, integrative and multidisciplinary assessments at various scales. It also requires the quantification of various uncertainties in both data and models, and the incorporation of stakeholders participation and institutional mechanisms into the various tools and risk assessment methodologies, to help decision makers understand and evaluate alternative measures and decisions. This requirement has been explicitly recognized in the context of the European Union Water Framework Directive (WFD). Designing the Programme of Measures is one of the key elements in building up the river basin management plan as defined by the WFD. The planning process for producing the Programme of Measures consists of several consecutive steps, starting from setting the objectives to the analysis of alternative ways of reaching them. It becomes evident that this is a challenging task which needs support from appropriate information systems and modelling tools that are able to cope with the complexity of the water system and planning process. This is particularly important in the context of current discussions on climate change impacts on the water sectors and actions needed to adapt to those changes. Since changes in the availability of water resources and occurrence of extreme events will affect not only the water supply and sanitation services but

© 2011 IWA Publishing *Decision Support for Water Framework Directive Implementation* Volume 3. Edited by Peter Vanrolleghem. ISBN: 9781843393276. Published by IWA Publishing, London, UK.

also other economic sectors, like agriculture, forestry, tourism, inland navigation, hydropower, health, etc. it is important to prepare a framework which allows policy makers to make integrated assessments.

To respond to this challenge, the European Commission initiated a large CatchMod initiative in the context of the 5th EU Framework Programme for Research and Technological Development (FP5). Its objective was to develop and harmonize integrated modelling tools that could be used by water authorities for planning and managing water resources in an integrated way at the scale of river basins. This very ambitious initiative with long-term objectives has been implemented through a dedicated cluster constituted by a core group of several research projects and coordinated by a concerted action, the Harmoni-CA project. CatchMod and Harmoni-CA addressed several key issues in modelling research, including model concepts and regimes, scale issues and data availability, understanding stakeholders participation, participatory approaches and institutional mechanisms and their incorporation in the various biophysical models and risk assessment methodologies to help decision-makers understand and evaluate alternative approaches, etc. Science-policy integration is another important challenge that has been addressed by the CatchMod initiative and the Harmoni-CA project.

In many instances, the lack of communication and of a clear coordination mechanism hinders the research outputs from being used or simply known by policy makers, and hinders policy research needs from being communicated to the scientific communities in a timely fashion. By initiating and facilitating the discussion between the Common Implementation Strategy Coordination Group of the WFD and the various researchers, Harmoni-CA helped to foster synergies between researchers, policy makers and water managers on different levels to achieve a common understanding of the various WFD implementation issues. It also helped consolidate knowledge generated by several EC projects (of the CatchMod cluster and beyond) on various cross-cutting and complementary research issues, with a view of providing technical support for the implementation of the WFD.

This publication is one of the concrete outcomes of the Harmoni-CA concerted action as part of a 4-volume series of Guidance Reports that guide water professionals through the implementation process of the Water Framework Directive, with a focus on the use of ICT-tools (and in particular modelling). They are supposed to be complementary to the Guidance Documents produced by the EU Directorate General for Environment. An initial volume "**Water Framework Directive: Model supported Implementation - A Water Manager's Guide**", edited by Fred Hattermann and Zbigniew W Kundzewicz, provides a practical guidance to water managers and other interested stakeholders on the model-supported implementation of the WFD. The three other volumes bring more detailed guidance regarding a range of aspects of such WFD-implementation. This particular volume 3 focuses on the "**Decision support for WFD implementation**", containing two Guidance Reports: (i) model-based tools for evaluating the complex interaction between water management and agriculture, and (ii) decisions support systems for water resources management, in particular specifying guidelines for the development of new decision support tools. The two other volumes cover "**Modelling aspects of WFD implementation**" in four additional Guidance Reports and "**Integrated Assessment for WFD implementation: Data, economic and human dimension**" with five final Guidance Reports. These Guidance Reports were written as part of the work conducted within the different CatchMod cluster projects. Therefore, some of these reports were already finalized in 2005-2006 and report the status at the time. Evidently, research and practice have moved forward, but it is expected that the review and synthesis nature of the Guidance Reports makes that the information they contain is still very actual.

These four volumes provide a good example of disseminating research results of policy relevance. We hope that this publication will reach a broad audience, both science and practice oriented, thus helping to further strengthen the links between scientific research, policy and decision making.

Dr. Panagiotis Balabanis
Principal Scientific Officer
Directorate General for Research
Commission of European Communities

Prof. Peter A. Vanrolleghem, Editor
Current address: modelEAU
Université Laval
Québec (QC) Canada

Guidance Report III.1

Review of model-based tools with regard to the interaction of water management and agriculture

Guido M. Bazzani

Abstract

This guidance report focuses on over fifty available tools that have a wider application, have been developed and applied in the context of EU-funded projects, or that present specific aspects of interest to support the implementation of the Water Framework Directive (WFD). The review clarifies the policy context of European water management and agriculture, paying particular attention to the recent reform of the Common Agricultural Policy (CAP) and its interaction with the WFD. In order to categorise and summarise the various models and tools related to water management and agriculture, a framework based on four critical dimensions (spatial scale, irrigation measures, agricultural measures and economic analysis) is adopted.

The tools are categorised into three classes, namely agronomic tools, hydrology and water quality tools and economic and land use tools. Further information is provided, referring interested readers to a dedicated web portal that offers comprehensive, European-wide shared information. Moreover, this report also aims to give an insight into the ways in which transparency in decision-making is improved by models and tools, how they support current water managers' tasks and methods to compare different management options in terms of their ecological, economic and social impact. To this end, three examples of good application are presented. Finally, the review considers how models and tools should be used to support the implementation process of the WFD, particularly considering its demands on public participation.

Keywords

Information and Communication Tools (ICT), Models, Agriculture, Water, Common Agricultural Policy (CAP), Water Framework Directive (WFD)

 Decision Support for Water Framework Directive Implementation Volume 3. Edited by Peter Vanrolleghem. ISBN: 9781843393276. Published by IWA Publishing, London, UK.

1. LIST OF ABBREVIATIONS

ABM	Averting Behaviour Models
AES	Agro-Environment Schemes
BMPs	Best Management Practices
CAP	Common Agricultural Policy
CR	Contingent Ranking
CSIRO	Commonwealth Scientific and Industrial Research Organization
CV	Contingent Valuation
DPSIR	Driving force–Pressure–State–Impact–Response
DS	Decision Support
DSS	Decision Support Systems
EEA	European Environment Agency
EU	European Union
FADN	The Farm Accountancy Data Network
FAO	Food and Agriculture Organization of the United Nations
GFP	Good Farming Practices
GIS	Geographic Information Systems
GUI	Graphical User Interface
HPM	Hedonic Pricing Models
ICT	Information and Communication Technology
IT	Information Technology
K	Potash (potassium)
MCA	Multi-Criteria Analysis
MCDA	Multi-Criteria Decision Analysis
N	Nitrogen
NGO	Non-Governmental Organization
P	Phosphorous
PC	Personal Computer
PEC	PEsticides Concentration
RB	River Basin
RBMP	River Basin Management Plans
RD	Rural Development
RDM	Recreational Demand Methods
RUSLE	Revised Universal Soil Loss Equation
SDSS	Spatial Decision Support Systems
SFP	Single Farm Payment
SPS	Single Payment Scheme
TEU	Treaty of the European Community
TMDLs	Total Maximum Daily Loads

USDA-ARS	United States Department of Agriculture-Agriculture Research Service
USEPA	United States Environmental Protection Agency
USLE	Universal Soil Loss Equation
WFD	Water Framework Directive

2. EXECUTIVE SUMMARY

2.1. Objective

This guidance report is primarily directed at water and agricultural managers who wish to find out about agricultural policy reform and existing models and tools that support joint water and agricultural management processes. It may also be beneficial to water authorities, water irrigation and reclamation boards, farmers' associations and NGOs. The goals of the document are manifold:

- to clarify the policy context, paying special attention to the recent reform of the Common Agricultural Policy (CAP) and its interaction with the Water Framework Directive (WFD);
- to categorize existing tools and models related to water and agriculture;
- to encourage stakeholders to use models and tools by offering examples of good application;
- to suggest how models and tools should be used to support a participatory implementation process of the WFD in agriculture.

2.2. Interaction between the Common Agricultural Policy and the Water Framework Directive

Agriculture plays a central role in water management, since it is a significant utilizer of water resources in Europe, accounting for around 30% of total use. In southern Europe, water is a fundamental agricultural input, with irrigation accounting for over 50% of demand. Furthermore, agriculture and forestry, which cover more than three-quarters of the area of the European Union, play a key role in determining the rural economy and environmental quality. Environmental pressures due to agricultural activities, including water pollution, are often viewed as a serious problem in many areas.

Due to an agreement on the Mid-Term Review of the CAP reached in June 2003, the Common Agricultural Policy (CAP) has adopted a new model for European agriculture. This model reflects the multifunctional role played by farming, integrating economic viability, food safety, social balance and environmental concerns. The key elements of the reform, which completely changes the way in which the EU supports its farm sector, are: adoption of a "single farm payment" for EU farmers – aid paid to producers will no longer be dependent on type of production, thus "decoupling" aid from, or removing the causes of, indirect environmental damage; enforcement of "cross-compliance", which makes direct payments conditional on specific requirements; strengthening Rural Development (RD) policy, into which instruments directly concerned with environmental outcomes are inserted.

The WFD and RD policy both adopt planning processes. However, important differences arise between them, of which scale and time could cause major problems. Win-win situations for both policies require close attention in the implementation phase; representatives from different authorities should interact to create a common vision of agricultural and environmental problems.

2.3. Categorization of existing tools and models

This guidance report focuses on over fifty available tools that have a wider application, have been developed and applied in the context of EU-funded projects, or that present specific aspects of interest to support the implementation of the WFD. To categorize and summarize these tools, a framework based on four critical dimensions (spatial scale, irrigation measures, agricultural measures and economic analysis) is adopted. The tools are categorised into three classes:

- agronomic tools;
- hydrology and water quality tools;
- economic and land use tools.

This review demonstrates how most tools tend to shift towards a new class, embracing holistic integrated Decision Support Systems (DSS). This follows the observed trend in modelling, which is changing from monodisciplinary to multidisciplinary approaches, in which several domains are integrated.

The non-exhaustive list can support the identification of models and tools to address specific issues, including the correct consideration of domains, scales and types of measures. A table in the annex summarizes the relevant information and includes each tool's website and contact details.

Further information is provided, referring interested readers to a dedicated web portal that offers comprehensive, shared European information, such as WISE, the 'Water Information System for Europe'. Since innovation in computational technology will further enhance the frontier of modelling, web portals are a key instrument for water and agricultural managers to access updated information.

2.4. Examples of good application

Three examples of good application are presented. The selected case studies have been carried out recently in different European water basins in the context of EU-funded projects. The projects are:

- WaterStrategyMan in the ARID cluster
- the German Elbe River Basin
- the Piave River Basin in Italy.

The case studies give an insight into relevant questions such as: How can tools improve transparency in decision-making? How can tools support water managers in their present tasks? How can various management options be compared in terms of their ecological, economic and social impact?

2.5. Models and tools to support the participatory implementation process of the WFD

The final section of the guidance report explores considerations of how models and tools should be used to support the implementation process of the WFD.

Good practice in model application deserves continual attention. In fact, the credibility and impact of the information and insight that modelling aims to generate are highly dependent on the quality of the modelling exercise.

The political process-oriented nature of water management requires an adaptive approach, which should also consider how models and tools should be used. A multilevel approach in modelling is recommended. At a higher level, conceptual models can support the definition of a common conceptual framework for cross-disciplinary work involving authorities, managers, stakeholders and researchers. At lower levels, quantitative models can support specific analyses exploring the strength of interactions and the sensitivity of the system to changes. The information produced, clarifying the areas of greatest uncertainty and influence on system evolution under different assumptions, can help prioritize field research and support water and agricultural management. However, it cannot offer definitive solutions, since uncertainty and subjectivity are intrinsic to all decision processes, which tools can reduce but not eliminate.

Since water pollution caused by agricultural activities is not specific to irrigated agriculture, the design of policies capable of increasing water quality, while preserving the economic and social sustainability of agricultural systems, requires a clear understanding of the complex relation at different scales from the field/farm up to region catchments and basins. Such understanding can be favoured by a combination of existing tools used by multidisciplinary teams of well-trained people with adequate data in the participatory process.

The definition of packages of integrated measures addressing quality and quantity issues can be supported by tools. However, there is no general solution, since measures are case-dependent. Various tools include specific routines to support the economic analysis requested by the WFD in one or more of the addressed aspects, ranging from the economic analysis of water use, cost recovery and pricing schemes, up to the evaluation of benefits to the environment and society, and eventually of disproportionate cost. Nonetheless, the support provided by such tools is highly dependent on how they are used – at present, it is incumbent on the Member States to decide how such analyses and valuations should be carried out; no common guidelines are defined.

The translation of the experience gained in research to the management domain will probably be a long process, in which support by the scientific community should be considered essential. In order to bridge the gap to real-world decision processes, the following recommendations are deemed important:

- a multilevel approach in modelling that integrates conceptual and quantitative models should be adopted;
- participatory modelling involving end users and stakeholders in modelling exercises should be implemented to enable them to understand and review the various assumptions and their implications for the modelling results;
- training and capacity-building activities are necessary at the local level;
- models and tools need to be adapted to local pre-established approaches, not vice versa;
- models and tools should be viewed as Information and Communication Technology (ICT) components of a structured and flexible approach.

3. INTRODUCTION

This guidance report follows the 2nd Policy Workshop of Harmoni-CA/WP5: Interaction of the Common Agricultural Policy and the Implementation of the Water Framework Directive at European and Regional Levels, which was held on 4-5 April 2005 in Brussels. The Harmoni-CA Document: HCA-WP5-2005-Re05/ Final Version-Deliverable No. D 5.3.2 20.04.05 by Ilke Borowski and Johannes Heeb summarizes: "*The workshop gave 30 agricultural and water managers from European, national and regional authorities the chance to get their hands on tools/models which may support their management activities during the implementation of the European Water Framework Directive and the Common Agricultural Policy. During the workshop an intensive exchange between the developers of the presented tools/models took place. Seven existing and easily accessible tools/models were presented and evaluated according to management questions which were defined by the participants with regard to the changing conditions under the WFD and the CAP. . . . During the workshop the discussion of the seven presented tools/models showed that most of the tools/models can support the joint water and agricultural management processes in some regard. However there was no tool/model identified which was considered fully capable to sufficiently answer all questions raised by the policy makers with regards to the interaction of CAP and WFD. Also, no combination of the presented tools/models was able to do so. The main gaps of the tools/models were identified in terms of the general integration of different domains like water quality, quantity, ecology but especially of economic evaluation means*" (Borowski and Heeb, 2005, p. 3).

The guidance report includes a to-do list for tool/model developers, which contains a summary of the demands identified during the workshop from the policy side:

1. "*Supporting participation - The existing tools/models can support and facilitate participation. But stakeholders have to be involved more intensively in the tool/model development process to bridge the knowledge gap between them, the "river basin managers" and the scientists (or in other words to make better use of existing knowledge).*
2. *Supporting economic assessments - There is a clear need for tools/models to support integrated economic assessments of policy measures. Existing tools/models should be integrated with existing agro-economic tools/models.*
3. *Transferability and applicability - Models should be developed to make a region-specific evaluation of policy measures possible. This, however, means that rather than having a few general tools/models, there is a need for a large number of specific, regional tools/models; a need which requires more resources. Tools/models should facilitate conflict and synergy management between CAP and WFD.*
4. *Integration - Tools/models should be developed to make more systemic work possible, integrating ecological, economical and social issues at qualitative and quantitative levels.*
5. *Simulation of behaviour - Tools/models are needed to simulate farmers' behaviour in response to policy measures. This would help to work out concrete action plans and select the most sustainable development scenario based on the simulation results.*
6. *Motivate policy makers to use the tools/models - Tools/models are still mostly used in the academic world. In order to convince policy makers to*

use tools/models in their daily work, information about the abilities, the validity, the level of uncertainty and the conditions of applicability of the tools/models have to be clearer to the practitioners. Tools/models should make clear what added value they can provide to policy makers working in integrated and participator river basin management in the CAP – WFD context. Access to the tools/models should be made easier for policy makers: promising options are toolboxes as developed, e.g. in the BMW project or the Harmoni-CA toolbox (www.harmoni-ca.info). In addition to technical solutions, more work needs to be done to understand the working constraints of policy makers so that tools/models can be adapted to fit their work."

The difficulties that model-based tools have in supporting practical water management are further explored in Borowski and Hare (2007).

The existing gap between scientists and policy-makers is confirmed by Mysiak et al. (2006) (see Table 1).

The present guidance report offers a broader review on tools and models, addressing both water management and agricultural management issues. It also responds to the mission of Harmoni-CA to establish and maintain dialogue between tool/model developers and policy-makers to improve the use of tools/ models in management processes.

The report is organized as follows:

- The first section briefly analyses water use in agriculture, and endeavours to clarify the policy context, introducing the Common Agricultural Policy (CAP), the Water Framework Directive (WFD) and existing interactions between the two policies.
- The second section categorizes existing tools and models related to water management and agriculture.
- The third section describes three examples of good tool application.
- The final section considers how tools should be used to support a participatory implementation process of the WFD, and provides conclusions.

The guidance report includes fives annexes that address specific issues:

1. A list of EU-funded projects concerning tools for water and agriculture.
2. The CAP
3. Measures to implement the WFD and the CAP
4. Monetary valuation

An information sheet on tools and models for water management in agriculture

Table 1 Gap between scientists, policy-makers and water managers (adapted from Mysiak et al., 2006)

Scientists	Policy-makers	Water managers
Knowledge among modellers and Decision Support System developers on the demands made on them by the WFD is often limited	Policy-makers wish to use models but often have a significant lack of confidence in them (black boxes)	Water managers need tools that reflect their national or federal river basin planning policies
Developing new models or adapting existing ones for reuse is usually time–consuming, and therefore expensive	DSS tools are usually perceived as tools that potentially limit decision-making capabilities	
Funding mechanisms do not facilitate long-term efforts, nor the efficient integration and exploitation of existing knowledge	Policy-makers have limited time; they are put under pressure by deadlines imposed by the implementation of the WFD	
	At least empirical cognitive models are always available in competent administrations, and are used for decision-making	

4. THE WATER FRAMEWORK DIRECTIVE AND THE COMMON AGRICULTURAL POLICY

4.1. Water use and agriculture

> Key message:
>
> Agriculture is a significant user of water resources in Europe, particularly in Southern Europe, where it accounts for around 45-81% of total consumption and where water is a fundamental agricultural input. Water pollution due to agricultural activities is an important indirect effect in many areas, although it is not specific to irrigated agriculture. For this reason, the design of policies capable of increasing water quality, while preserving the economic and social sustainability of agricultural systems, requires a clear understanding of the existing complex relation, which is always site-specific.
>
> Conceptual models are important tools to identify such relations. They also favour interdisciplinary dialogue, enabling experts, stakeholders and decision-makers to find a common vision in order to define packages of integrative measures, addressing quality and quantity issues.

Agriculture is a significant user of water resources in Europe, accounting for around 30% of total use. This is particularly true in the Mediterranean region (including Greece, Italy, Portugal and Spain), where irrigated agriculture, although it covers only 20-25% of the total cultivated surface, accounts for about 50% of total agricultural production and 45-81% of the total water demand[1].

Furthermore, agriculture and forestry, which cover over three-quarters of the area of the European Union, play a key role in determining the rural economy and the environment.

Agricultural systems changed radically in the second half of the previous century. Diversified evolution patterns emerged and unforeseen negative consequences developed over time. On the one side, specialization of holdings took place in both livestock and arable sectors, based on farm enlargement and capital "intensive" farming methods, which exerted great pressure on the environment; on the other side, decline and abandonment occurred.

A variety of driving forces that underlie farmers' management decisions influenced this process: technological development, changes in market conditions, alterations in the costs of labour, land and other factors of production, modification

[1]Source: http://www.fao.org/nr/water/aquastat/data/query/index.html

of the socio-cultural environment, structural changes and a range of different public policies, including the CAP itself.

In the market economy, the comparative profitability of farming, which is highly influenced by the CAP, forced farmers to pursue financial objectives. Consequentially, traditional practices were often abandoned in favour of intensive ones: the substitution of labour with capital and of organic with industrial inputs increased farm productivity at the cost of environmental sustainability. As a consequence, water pollution due to agricultural activities is often considered a serious problem in many areas.

However, agriculture is not a homogeneous sector. In most regions, farms that use intensive management practices, exerting great environmental pressure, coexist with others that are environmentally friendly. In many cases, the latter group is dwindling in size, although only marginally in economic terms. This implies the need for targeted interventions in order to focus on appropriate management at the farm level. This would lead to favouring long-term sustainability through the balanced development of rural areas.

Devising policies that are not only capable of reducing consumption and increasing water quality but also of preserving the economic and social sustainability of agricultural systems requires a clear understanding of the complex relation between water and agriculture. This task can be enhanced by constructing conceptual models that favour interdisciplinary dialogue and the definition of a common vision among experts, stakeholders and decision-makers. Estimating the effects of a policy measure requires the identification of the causal links between the implementation of the measure and its ultimate impact on human activities and the environment. A conceptual framework for examining such links can be found in the DPSIR (Driving force–Pressure–State–Impact–Response) approach proposed by the European Environment Agency (EEA) for the description of environmental pathways. In a policy-making context, this approach can help policy-makers to conceptualize and structure their decisions on potential alternative measures, according to the cause-effect relationships.

Figure 1 is an example of a conceptual model of water and agriculture.

- The model identifies the qualitative and quantitative dimensions of water.
- Point and diffuse sources of pollution are connected with agriculture and farming, although they are not specific to irrigated agriculture; there is strong evidence that in many regions rain-fed agriculture and intensive animal breeding without land are the main causes of environmental impact, e.g. nitrates and other pollutants.
- Concerning agricultural irrigation, it is important to understand how farmers use water and make decisions in response to external stimuli. From

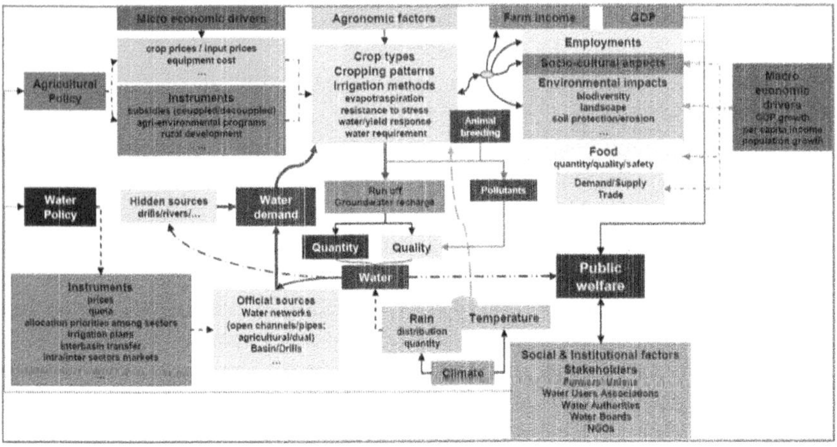

Figure 1 Example of a conceptual model of water management in the agricultural system

an economic perspective, water is a production factor, which substantially enlarges farmers' sets of choices in terms of available crops and processes (Ward and Michelsen, 2002). Irrigation has other important effects, such as the higher quality of production and risk reduction due to uncertain and unstable climate conditions, particularly for fruit, vegetables and other high-value crops. Furthermore, water enables farmers to standardize production over space and time, which is becoming a stringent market requirement, while the quantitative increase of production is merely a secondary effect.

- Water demand in agriculture is linked to agronomic factors (Doorenbos et al. 1979). On the one hand, crops have a different effect on evapotranspiration, resistance to stress and water/yield response. On the other hand, agricultural practices, cropping patterns and irrigation methods can significantly influence water requirements at the farm level and in terms of soil characteristics.
- Climatic conditions impact on water demand and availability in many ways. For instance, temperature influences plants' evapotranspiration, water needs and rain distribution, and quantity is a key factor regarding water availability and water quality, since scarcity is generally linked to a higher concentration of pollutants in the resource.
- Economic factors operate at both macro and micro levels. At the macro level, the rate of growth of gross domestic product (GDP), the distribution of income in society and population growth are key drivers for demand in food,

which can be differentiated into quality and quantity, and into perceptions of and the demands placed on the rural environment. For instance, demand for recreation in rural areas, and hence for environmental quality, is higher in rich societies where basic needs have already been satisfied. At the micro level, prices and costs at the farm level, which vary between crops and farms, as well as investment and operating costs for irrigation equipment influence farmers' choices.

- Social and institutional factors play a central role in water management and should be considered with due care. All stakeholders represented by Farmers' Unions, Water Users' Associations, Water Authorities, Water Boards and NGOs should be involved in the participatory process of the implementing the WFD.
- Policy factors, which represent a response of society to impact, play a wide range of roles. There are many instruments that model water and agriculture (see Section 2.2).
- Agriculture is a multifunctional activity that should be assessed in a multicriteria framework. Relevant dimensions are:
  - the economic dimension: quantified at the micro level by the income produced by farm type and at the macro level by the contribution of the aggregate sector to GDP.
  - the social dimension: represented not only by employment induced but also by the cultural aspects linked to rural life,
  - the environmental dimension, which is particularly complex: positive and negative indirect effects are both present. On the one hand, positive 'externalities' are that appreciated landscapes[2] depend on water availability in agriculture, biodiversity conservation and soil protection. On the other hand, soil erosion, soil and water pollution and subsidence sometimes pose serious problems.
  - the dimension of food production: this traditional objective of agricultural activity still plays an important role. This needs to be assessed properly by considering the high value of European production with reference to food quality and safety, due to the strict rules enforced by the CAP. For example, genetically modified crops (GMO crops) cannot be cultivated due to precautions; there are limits to pesticide concentrations on fruit and vegetables, which are much more stringent for organic farming, while imported crops and food do not follow the same requirements.

[2]Irrigation networks are often used to drain excess water in the rainy season, which is an important environmental service.

- Water distribution networks deserve specific attention. Many water distribution networks are old, operating under very poor conditions with low maintenance, if at all. In such cases, high losses (up to 50%) are frequent. Measures focusing on water management at the farm level cannot address this problem. However, water savings could be achieved by investing in infrastructure.
- Demand should be satisfied using water derived from legal sources, such as irrigation networks and authorised basins, drills and rivers. However, hidden sources are a real option in many situations. The use of hidden water as an alternative to legal sources is particularly relevant, considering that economic instruments (e.g. water prices, water quotas, etc.) can only be applied to the latter, and the cost of enforcing an effective monitoring policy to reduce illegal behaviour could be extremely high.
- The groundwater recharge effects of agricultural irrigation are another critical issue to be considered. If water is not polluted, runoff and percolation are a positive externality. In such cases, the environmental benefits induced by an increase in irrigation efficiency (e.g. due to a shift from surface to sprinkler irrigation) or the reduction of irrigation itself may be questionable.

4.2. The Common Agricultural Policy reform

> Key message:
>
> The 2003 CAP reform should secure multifunctional, sustainable and competitive agriculture throughout Europe.
>
> The reform reduces incentives to produce intensively by "decoupling" payments from production. Moreover, the adoption of "cross-compliance" as a mandatory principle incorporates basic standards for the environment, food safety, animal health and welfare in common market organizations, as well as the maintenance of farms in good agricultural and environmental conditions.
>
> The rural Development policy, which offers direct opportunities to improve environmental quality, is strengthened.

Since its establishment, the whole concept of the CAP and its method of operationalisation have greatly changed through successive reforms. The integration of environmental issues into other EU policies, such as the CAP, is part of a long process dating back to the 1980s. It is legally bound in the Treaty of the European Community (TEU): "*Environmental protection requirements must be integrated into the definition and the implementation of (all) the Community policies and activities . . . in particular with a view to promoting sustainable development*[3]". The Agreement on the Mid-Term Review of the CAP, reached in June 2003, gives form to a new European model for agriculture, reflecting the multifunctional[4] role that farming plays, and integrating economic viability, food safety, social balance and environmental concerns. The key elements of the reform, which completely changes the way in which the EU supports its farming sector, are:

1. adoption of a "Single Farm Payment" (SFP) for EU farmers: aid paid to producers will no longer be dependent on type of production, thus "decoupling" it from or removing the causes of indirect environmental damage[5];
2. introduction of a "cross-compliance" rule, which makes direct payments conditional on adherence to specific requirements[6]; and
3. strengthening of the RD policy, in which instruments directly concerned with environmental outcomes are inserted[7].

The previous elements are described in further detail in Annex 2.

The new more decentralized model of agricultural policy grants Member States greater freedom by reducing the risk of distorting competition or renationalizing the CAP. The reform should secure multifunctional, sustainable and competitive agriculture throughout Europe[8].

[3]The 5th Environmental Action Programme refers to sustainable development as "development which meets the needs of the present without compromising the ability of future generations to meet their own needs". This entails preserving the overall balance and value of the natural capital stock and taking a long-term view of the real socio-economic costs and benefits of consumption and conservation. The achievement of "balanced and sustainable development", together with the promotion of economic and social progress and a high level of employment, represents the first overall objective of the Union (Art. 2 of the TEU).

[4]The term conveys that agriculture provides a variety of non-marketable outputs and services of good public character, which are valuable to the well-being of society.

[5]Member States have the option to decouple between 2005 and 2007; however, to deflate its impact and prevent the abandonment of production, states can maintain a limited link between subsidies and production under well-defined conditions and within clear limits.

[6]The SFP is linked to regulations in the fields of the environment, public, animal and plant health and animal welfare; farmers are sanctioned for non-compliance (partial or entire reduction of direct support). Beneficiaries of direct payments are also obliged to keep their land in good agricultural and environmental condition.

[7]RD forms the "second pillar". The shift from the first to the second pillar is strengthened by another innovative measure: "modulation", which makes an additional RD fund available, thanks to a 5% reduction in direct payments for bigger farms.

[8]This is predicted to have a significant impact on the European farm sector. The EU will probably lose most of its share of the world export market for dairy products. Exports of coarse grains are expected to fall. The EU will also import far more beef and poultry.

4.3 The Water Framework Directive

Key message

The Water Framework Directive (WFD) establishes a framework for common action in the field of water policy in Europe. Its main goal is to achieve a good water status for nearly all European waters by the year 2015.

The WFD recommends economic approaches and tools and the consideration of economic instruments in order to achieve its environmental objectives in the most effective manner. Different aspects are addressed, ranging from the economic analysis of water use, cost recovery and pricing schemes, to the evaluation of benefits to the environment and society, and eventually to disproportionate costs.

At present, Member States are left to decide how such analyses and valuations should be carried out; no common guidelines have been defined.

The 2000/60/EC Directive, known as the Water Framework Directive (WFD), establishes a framework for common action in the field of water policy in Europe. Its main goal is to raise the water quality of nearly all bodies of water to a "good status" by 2015. The directive is an important element of European environmental policy.

In the 1990s, economic instruments were increasingly suggested and recommended to enhance the sustainability of the environment. They found their full legitimacy in the Rio Declaration on Environment and Development by the United Nations in 1992. The WFD has a similar emphasis. In fact, it calls for the application of economic principles, i.e. the polluter pays principle, and cost recovery for water services, with the inclusion of environmental and resource costs. The directive clearly recommends economic approaches and tools, and the consideration of economic instruments to achieve its environmental objectives in the most effective manner.

The WFD identifies a strict time schedule for key tasks aimed to develop and implement River Basin Management Plans (RBMP). Moreover, article 5 of the WFD requires that each Member State shall ensure that an economic analysis of water use is undertaken for each river basin district or part of an international river basin district falling within its territory, according to the technical specifications set out in Annexes II and III. This should be completed within four years of the Directive coming into force.

The principle of the cost recovery (Art. 9) of water services, including supply, environmental and resource costs[9] should be adopted in accordance with the polluter pays principle.

Furthermore, Art. 9 requires that Member States shall ensure by 2010 that water pricing policies provide adequate incentives for water users to use water resources efficiently, thereby contributing to the environmental objectives of the WFD.

Economic analysis can also justify derogation, including the designation of a water body's status, (Art. 4) if disproportionate costs can be demonstrated. The affected groups' ability to pay should be considered in the cost assessment.

What emerges from the previous articles is that different aspects are addressed, ranging from the economic analysis of water use, cost recovery and pricing schemes, to the evaluation of benefits to the environment and society, and eventually disproportionate costs.

The following activities, which involve economic components, should be undertaken in the implementation process:

- An economic analysis of water use with the main objective of assessing the importance of water to the economy and the socio-economic development of river basin districts (RBD). Tradeoffs between socio-economic development and water protection should be estimated and made available.
- The dynamics of the RBD should be adequately investigated and the baseline scenario defined. Forecasts of the key economic drivers likely to influence water pressure should be made and the water status should be given.
- An assessment of the current levels of cost recovery for water services, considering financial, environmental and resource components, and how they are distributed among water users.
- A cost-effectiveness analysis of the main measures to be considered in the development of the RBMP.

The economic valuation of benefits is a challenging task, since benefits are many and very different in nature. A reduction of water treatment costs is a market

[9]Definitions are based on the WATECO (2003) glossary.

- The supply costs (direct or financial costs) of water services include the costs of providing and administering these services. They include all operation and maintenance costs, and capital costs (principal and interest payment), and return on equity, if appropriate (p.70).
- Environmental costs are the costs of damage that water use causes to the environment and ecosystems, and the costs of those who use the environment (e.g. a reduction in the ecological quality of aquatic ecosystems; salinization and degradation of productive soils). (p.69)
- Resource costs: costs of foregone opportunities that other uses suffer from due to the depletion of the resource beyond its natural rate of recharge or recovery (e.g. linked to the over-abstraction of groundwater (p.72)).

benefit that can easily be valued. Since the goal of the WFD is to achieve higher water quality, which translates into more environmental services, many benefits are of a non-market nature. Non-market benefits could be linked to use, such as an increase in open-access recreation. They may also be values that individuals do not relate to higher environmental quality.

Until now, Member States are left to operationalize the form of analyses and valuations. No common rules or methodologies have yet been defined. Hence, the Member States are free to adopt the procedure they wish to apply.

For further information on economic valuation, see Annex 4.

Finally, the WFD also requires an integrated participatory water resources policy, which high-quality computer-based tools could support.

4.4 The CAP and the WFD

Key message.

The WFD and the CAP need to be integrated. However, differences in objectives, scale and time-table suggest that the implementation phase should receive greater attention.

Representatives from the different authorities should interact to define programmes of measures that can create a common ground for agricultural and environmental policies.

Both the WFD and the RD adopt planning processes. However, important differences emerge, of which scale could pose a major problem: the WFD identifies River Basins (RB) as the proper planning scale, whereas the CAP follows a national/regional approach[10]. Win-win situations for both policies require that great attention is paid in the implementation phase. They also require representatives from different authorities to interact, in order for them to construct a common vision of agricultural and environmental problems.

The first column of Table 2 shows different policy objectives of the WFD: the quality objective identified in the "good water status"; the quantity objective related to a reduction in water demand and the economic objective related to the recovery of costs for services. Specific policy instruments are identified in the second column: some are compulsory and others are not, some belong to agricultural policy (A) and others to water policy (W), others still are a combination of both. The main effects are reported in the final column.

[10]The different timetables stipulated by the WFD and the CAP create further difficulties.

Table 2 WFD policy objectives, agro-environmental instruments and possible effects on water and agriculture (A) Agricultural Policy; (W) Water Policy

Policy objectives	Instruments		Possible effects
Good water status Dir. 91/271 discharge Dir. 91/676 nitrates	Zoning	A	
	Decoupling subsidies	A	
	Agro-environmental schemes	A	
	Cross-compliance measures (standard)	A	
	Voluntary agreements (higher quality)	A	
	Tax on pollutant input	A/W	
	Markets for pollution rights - tradable permits	W	
	. . .		
Water saving	Decoupling subsidies	A	Higher water quality
	Cross-compliance measures (mandatory)	A	Water savings
	Voluntary agreements (higher reduction)	A	Land use changes
	Improved irrigation distribution network	A	Different cropping patterns
	Innovation in irrigation technology	A	Low agricultural employment
	Water rights	W	Farm income reductions
	Planning water allocation in space and time	W	Transaction costs
	Water transfer infrastructure	W	Distributional effects
	Water prices (only metering)	W	Social welfare variations
	Abstraction licences and quotas plus fees	W	
	Water markets	W	
	Alternative sources, including reuse of waste water (considering the quality/use relation)	W	
	Small basins	A/W	
	. . .		
Cost recovery	Water pricing		
	General taxation		

Many of the previous instruments require the definition of specific measures (for more detailed information on agricultural measures, see Annex 3). Many agricultural measures aim to wholly or partly improve or protect water quality (e.g. agricultural measures to reduce the use of pesticides and fertilizers) or water quantity (e.g. measures to reduce irrigation). Specific actions and measures could involve land use activities, requiring changes in land use and management (e.g. development of low-input farming systems, such as organic farming, changing from arable to grassland, investments to improve the state of irrigation infrastructure and allowing farmers to shift to improved irrigation techniques). Some instruments, such as buffer strips, crop diversity and rotation, are agronomic in nature; others, such as taxes on pollutant input, are economically-oriented.

Past experience suggests the adoption of an approach that combines schemes. Such an approach would integrate broad and shallow measures seeking widespread basic services with other narrow and deep measures that pursue clearly defined, more demanding, targeted benefits in particular priority areas.

When combined schemes are adopted, given the systemic nature of the environment, it is impossible to isolate the specific contribution of a specific measure to a given objective; instead, multicriteria valuation is used to identify the advantages and disadvantages of alternative schemes. The WFD requires that the cost-effectiveness of such alternatives should be considered, which implies rather complex economic analysis.

It is essential to integrate and coordinate between the two policies at the implementation level, in order to reduce or avoid the risk of conflict, and to take advantage of the potential of RD programmes to deliver the WFD's objectives. Future RD budgets suggest that some kind of prioritisation of objectives will be necessary, since funds are limited. On the other hand, close attention should be paid to the design of the RBMP, since it can also have a huge impact on RD. In this respect, the WFD comprises tools that can be used to mitigate conflicts: public participation, economic analyses and derogations. Alternatives should be considered when the most cost-effective measures involve "disproportionate costs" for rural communities, which could put the sustainability of agricultural activity at risk.

5. CATEGORISATION OF EXISTING TOOLS AND MODELS RELATED TO WATER AND AGRICULTURE

Key message:

This review focuses on over fifty available tools that have a wider application, have been developed and applied in the context of EU-funded projects, or that present specific aspects of interest to upport the implementation of the WFD. To categorize and summarize these tools, a framework based on four critical dimensions (spatial scale, the implementation of irrigation and agricultural measures, economic analysis) is adopted. The tools are categorized into three classes:

- agronomic tools
- hydrology and water quality tools
- economic and land use tools

This review shows how most tools tend to shift towards a new class, embracing holistic integrated Decision Support Systems. This follows the observed trend in modelling, which is developing from monodisciplinary towards multidisciplinary approaches, in which several domains are integrated.

The non-exhaustive list can support the identification of models and tools to address specific issues, including the proper consideration of domains, scales and types of measures. A table in the annex summarizes the relevant information and includes each tool's website and contact details.

Further information is given, referring interested readers to dedicated web portals that offer comprehensive, shared European information, such as WISE, the 'Water Information System for Europe'. Since innovation in computational technology will further enhance the frontier of modelling, web portals are a key instrument for water and agricultural managers to access updated information.

There are so many tools related to water and agriculture that it is virtually impossible to offer an exhaustive review. The approach adopted in this guidance report is to focus on available tools that have a wider application, that have been developed and applied in the context of EU-funded projects or that provide specific aspects of interest to support the implementation of the WFD.

All of the investigated tools are summarized in the table shown in Annex 5, giving the following information: tool name, brief description, spatial scale, irrigation measures, agricultural measures, economic analysis, type of measures included, link to the tool's website, contact details, further characterizations and previous applications.

To categorize and summarize the tools, we adopt a framework based on four critical dimensions: spatial scale, followed by the implementation of irrigation and agricultural measures, and concluding with an economic analysis.

To synthesize the highly inhomogeneous information from the previous four dimensions, we have adopted a simplified classification approach. The spatial scale represents the geographic unit of the tool – in many cases, such a dimension is not unique; the predefined levels are: farm, catchment, sub-basin, basin, region, State and Europe. Two distinct dummies (YES/NO) identify the tool's capability to analyze irrigation and agricultural measures. A three-position scale is adopted to classify the economic analysis; the code reflects the economic approach adopted in the tool: MI = micro economic, MA = macro economic, OA = other approach.

Tools can embed one or more models, which are 'a simplified representation of a system (or process or theory) intended to enhance our ability to understand, predict, and possibly control the behaviour of the system' (Neelamkavil, 1988).

A few decades ago, tools were mainly monodisciplinary, used to describe relatively simple problems in specific domains of the economy, agronomy, ecology, hydrodynamics, groundwater or surface water quality. In the past decade, we gradually moved towards more multidisciplinary approaches to problem-solving, in which several domains are integrated.

Due to its demand to integrate groundwater, surface water, ecological and economic aspects of water management at the river basin scale (holistic approach), and due to the explicit requirement to study the impact of alternative measures (human interventions), the WFD seems to support this trend of exploiting more sophisticated, integrated models.

By following the adoption of the integrated water resources management paradigm, in which public participation is a key element, the WFD moreover explicitly calls for public participation and active stakeholder involvement in the process of water resources management, and hence also in the modelling process (Pahl-Wostl, 2002). There is widespread agreement that stakeholder involvement does not imply active participation in the technical modelling itself, but rather appears as a demand to be able to understand and review the various assumptions and their implications for the modelling results (Refsgaard, 2005).

The tools have been classified into three classes, considering the original core of the software:

- agronomic tools
- hydrology and water quality tools
- economic and land use tools

Most tools, however, are developing towards a new class, embracing holistic integrated DS.

5.1 Agronomic tools

This group contains agronomic models to study crop yield response to water and chemicals, plant protection models and tools to support the process of feed planning and budgeting. The observed trend in this group of models and tools is also towards stronger integration.

Such tools are even more relevant in agriculture-intensive areas, where the environmental impact of rural development plans, manure and fertilizer management and of agrotechniques involving the use of agrochemicals can be assessed.

There are many versions of software available to model, and therefore predict, the agronomical, environmental and economic consequences of the complex interactions between crop management, soil and the atmosphere. Few of these guarantee a simple approach; most utilize complex relationships, e.g. models that allow the simulation of stochastic scenarios (Acutis et al., 2000; Peralta and Stöckle, 2001), which can be useful in the estimation of probabilities associated with the occurrence of events. For further references, see: Donatelli et al., 1997; Johnsson et al., 2002; Lewis et al., 2003; Ten Berge et al., 2000; Wolf et al., 2003.

Many of these tools were developed in the United States of America. Examples include:

- The EPIC 'Erosion Productivity Impact Calculator' model (Williams 1990; Williams 1995) is a widely used simulation tool for agricultural policy analysis. The model, originally developed by the US Department of Agriculture (USDA), is now maintained by the Texas A&M Blacklands Research Center. EPIC is a field-scale model that can be adapted to a large range of crop rotations, management practices and environmental conditions. The model was designed to assess the impact of soil erosion on crop productivity (Williams et al., 1984). The new version is called 'Environmental Policy Integrated Climate' (Mitchell et al., 1996), reflecting the evolution of the tool. Exemplary applications include estimations of soil erosion from water and wind, and climate change impacts on crop yield and soil erosion. EPIC is also used as part of Agricultural Policy/Environmental eXtender APEX, a tool for managing whole farms or small watersheds to obtain maximum production efficiency and maintain environmental quality. Examples of its application include: terrace systems, grass waterways, strip cropping, buffer strips/vegetated filter strips, crop rotation, fertilizers, irrigation, liming, furrow diking, drainage and waste management (feed yards, dairies with or without lagoons).

- Another interesting tool is CropSyst (Stöckle et al., 2003), which is a cropping system simulation model, distributed free of charge. The model was developed as an analytic tool to study the effect of cropping system management on productivity and the environment. The model simulates the soil water budget, the soil-plant nitrogen budget, crop canopy and root growth, dry matter production, yield, residue production and decomposition and erosion. Management options include: cultivar selection, crop rotation (including fallow years), irrigation, nitrogen fertilization, tillage operations and residue management. A link to economic and risk analysis models is under development. Four input data files are required to run CropSyst: location, soil, crop and management files. The separation of files makes it easier to link CropSyst simulations to Geographical Information System (GIS)[11] software. CropSyst provides a platform for simulating crop rotations, an automatic management events scheduler and the possibility to run multiple simulations in connection with a GIS. All these characteristics make CropSyst ideal for scenario simulations; other models simulate crop growth processes with more detail but have lower or no flexibility in specifying routine management techniques.

- CREAMS 'Chemicals, runoff and erosion from agricultural management systems' is a field-scale model for predicting runoff, erosion and chemical transport from agricultural management systems. It is applicable to field-sized areas.
- GLEAMS, 'Groundwater Loading Effects of Agricultural Management Systems', can simulate edge-of-field and bottom-of-root-zone loadings of water, sediment, pesticides and plant nutrients from complex climate-soil-management interactions. The tool provides estimates of the impact of management systems (e.g. planting dates, cropping systems, irrigation scheduling and tillage operations) on the potential for chemical movement. GLEAMS can be useful in long-term simulations for pesticide screening in soil/management. It can track the movement of pesticides with percolated water, runoff and sediment. The upward movement of pesticides and plant uptake are simulated with evaporation and transpiration. Degradation into metabolites is also simulated for compounds that have potentially toxic bi-products. Erosion in overland flow areas is estimated using a modified

[11]A GIS is a computer system capable of capturing, storing, analysing and displaying geographically referenced information, that is, data identified according to location. Practitioners also define a GIS as including the procedures, operating personnel and spatial data that go into the system.

Universal Soil Loss Equation. Erosion in chemicals and deposition in temporary impoundments, such as tile outlet terraces, are used to determine sediment yield at the edge of the field[12].

- AGNPS, the 'Agricultural Non-Point Source Pollution Model', is a computer model developed jointly by the USDA Agricultural Research Service and the Natural Resources Conservation Service to predict non-point source pollutant loads within agricultural watersheds.

The integration of agronomic models with GIS reflects recent developments.

The 'Agricultural Production System Simulator', or APSIM, is another well-known tool to simulate biophysical processes in farming systems. The tool, developed in Australia by the Commonwealth Scientific and Industrial Research Organization (CSIRO), integrates models derived from fragmented research efforts, enabling comparisons to be made on a common platform. The system includes plant, soil and management modules. These modules consider a diverse range of crops, pastures and trees, soil processes including water balance, N and P transformations, soil pH[13], erosion and a full range of management controls. APSIM resulted from a need for tools that provide accurate predictions of crop production in relation to climate, genotype, soil and management, while addressing long-term resource management issues. It relates to the economic and ecological outcomes of management practices in the face of climate risk.

There are many interesting developments in Europe, including:

MicroLEIS DSS, 'A Land Evaluation Decision Support System for Agricultural Soil Protection', is a computer-based set of tools for the orderly arrangement and practical interpretation of land resources/agricultural management data, developed and applied in Spain. The DSS is an agro-ecological system. Its major characteristics are: data and knowledge engineering through the use of a variety of databases and innovative modelling techniques; the scaling-up of process knowledge from the micro scale to the landscape scale; land evaluation; use of monthly meteorological data and standard information; an integrated agro-ecological approach, combining biophysical data with agricultural management experience; software development for PC platforms, web- and GIS-based versions.

CROPWAT is a Decision Support System (DSS) used for irrigation planning and management, developed by the Land and Water Development Division of the FAO. It is intended to be a practical tool to help agro-meteorologists, agronomists and irrigation engineers perform standard calculations for evapotranspiration and

---

[12]Source: http://www.tifton.uga.edu/sewrl/Gleams/gleams_y2k_update.htm

[13]pH is a measure of the acidity or alkalinity of a solution.

crop water use studies and, more specifically, to devise and manage irrigation schemes. It allows the assessment of production under rain-fed conditions or deficit irrigation. Standard crop data are included in the program, and climatic data can be obtained for 144 countries through the CLIMWAT database. CLIMWAT is a climatic database that can be used in combination with the computer program CROPWAT. It allows the immediate calculation of crop water requirements, irrigation supply and irrigation scheduling for various crops for a range of climatological stations worldwide.

SIMIS, the 'Scheme Irrigation Management Information System', is a tool designed to facilitate the management tasks of irrigation schemes. This program, developed by the FAO, is not limited to water aspects, but covers all major issues of day-to day management activities, including the control of maintenance, accounting, water fees and other relevant tasks. FAO advice states that SIMIS should no longer be supported.

The creation of web systems to promote the dissemination of agronomic information is a promising element to support resource management. Unfortunately, however, the following interesting applications do not have an English version:

- PlanteInfo is an information and decision support system for farmers and agricultural advisers, developed and applied in Denmark. Most of the information is generated dynamically with models using frequently updated databases. A subscription system enables personalized information to be accessed. For example, since the geographic location of a user's home is known after login, local weather data are used for model calculations. PlanteInfo can store users' previously entered information, such as fields, crops and actions, for future use. PlanteInfo also has a public version, which contains the facilities visible in the menu.
- IRRINET is an example of agrometeorology that integrates data from different sources, up to personalized, guided irrigation scheduling for farmers. Guided irrigation scheduling is based on water balance models at the plot level, based on local weather data and information on specific agro-techniques. The service has been employed in Northern Italy since 1985, and yields an average 20% reduction in water consumption.

5.2 Hydrology and water quality tools

Models describing water flows, water quality and ecology are being developed and applied in increasing number and variety. The observed trend in the past decade is towards increasingly sophisticated computerised systems, integrating watershed processes that operate at different spatial and temporal scales, simulation models and decision-making approaches.

These tools have been developed for a variety of purposes, such as the prevention of water shortages (drought), surpluses (floods) and water impairment (pollution). The complexity of model-based water management has extended even further, as integrated system dynamics and stochastic simulations models, known as ecohydrological models, represent the state of the art of formal modelling (Reca et al., 2001). Ecohydrology combines the study of hydrological, biogeochemical and ecological processes and their interrelations in soil and water bodies. This type of model for a river catchment contains a hydrological module as its basic element and a vegetation sub-model, which usually includes further sub-models for biogeochemical cycles (carbon, nitrogen, phosphorus) with an appreciable level of complexity. The previous sub-models are usually combined to include interaction and feedback between processes, such as water and nutrient drivers for plant growth, water transpiration by plants, nutrient transport with water, etc.

Many earlier tools adopted nodal network approaches, which are a common framework for considering water allocation problems (see, for example, McKinney et al., 1999; Rosegrant et al., 2000; Merritt et al., 2004; Letcher et al., 2007; Jakeman and Letcher, 2003; Fedra and Jamieson, 1996). In this type of model framework, a river basin is represented as a series of nodes. Nodes represent points where extraction and other activities that impact the stream are aggregated and modelled for a region. Regions refer to land or users attached to a node. Depending on the problem addressed by the model, regions may be defined by physical boundaries (e.g. sub-catchment areas) or by social, economic, technical and political boundaries. One example of this type of boundary is the property areas of irrigators who extract along a reach of a stream between two nodes. Flows are generally routed from upstream nodes to downstream nodes. Thus the impact of upstream land and water use activities on downstream users is modelled.

Examples of rules associated with nodes in a hydrological model are given in Table 3.

The reliability of process-based models, their flexibility and level of integration have improved in the past decade (Krysanova et al., 2005; Quinn et al., 2004; Hattermann et al., 2004; He, 2003; Krysanova et al., 2007).

The United States Environmental Protection Agency (USEPA) supports and recommends that state and federal agencies use a set of models available within a framework called 'Better Assessment Science Integrating Point and Non-point Sources' (BASINS). Created in 1996 with subsequent releases in 1998, 2001, and 2004, the tool is a multipurpose environmental analysis system designed for use by agencies to conduct watershed and water quality-based studies. BASINS integrates environmental data, analytical tools and modelling

Table 3 Examples of rules associated with nodes in a hydrological model

Type	Purpose	Specification Required
Abstraction	Enforces a water user (water supply, irrigation, hydropower) to receive enough water to cover its demand, as given in the user's input time series	Upstream node on river as water source and downstream user node
Minimum flow	Enforces minimum flow at nodes	Relevant node on river (no downstream node), time series of flow requirement
Reservoir storage	Enforces storage in reservoirs up to flood control level	Relevant reservoir node (no downstream node, no time series)
Reservoir target level	Enforces water levels in reservoirs	Relevant reservoir node (no downstream node), time series of target levels
Specified abstraction	Enforces a water user to receive enough water to cover its demand, as given in a separate time series (overriding input time series)	Upstream node on river and downstream user node, time series of demand

programs to support the development of cost-effective approaches to watershed management and environmental protection, including 'Total Maximum Daily Loads' (TMDLs).

BASINS 4.0 and BASINS 3.1 represent two co-existing versions. In BASINs 3.1, the watershed loading/water quality embedded model is the 'Soil Water Assessment Tool' (SWAT) (Arnold et al., 1993; Di Luzio et al., 2002), developed by the United States Department of Agriculture-Agriculture Research Service (USDA-ARS). The SWAT model was created in an attempt to simulate processes as physically and realistically as possible. Most of the model inputs are physically based (that is, based on readily available information). It is important to note that SWAT is not a 'parametric model' with a formal optimisation procedure (as part of the calibration process) to fit any data. Instead, a few important variables that are not well defined physically, such as the runoff curve number and Universal Soil Loss Equation (USLE) cover and the management factor (such as the 'C factor', which quantifies the vulnerability of specific land use to water erosion), can be adjusted to provide a better fit. SWAT has been applied in numerous hydrological and/or non-point source pollution studies (http:// www.brc.tamus.edu/swat/swat-peer-reviewed.pdf), including in Europe.

BASINS 4.0 uses an open source GIS software architecture, which can easily be shared with other GIS software. BASINS 4.0 includes all of the functionalities of BASINS 3.1, except the AGWA[14] and SWAT models. Furthermore, it has the ability to quickly develop and create "plug-in" functions to update or enhance the watershed analysis process and monitor the environment. Both tools adopt QUAL2K, the 'River and Stream Water Quality Model', as their water quality model.

MIKE BASIN (MB) is a commercial, versatile GIS-based water resource and environmental modelling package produce by DHI[15]. It provides a simple, yet powerful framework for managers and stakeholders to address multi-sectoral allocation and discharge issues in a river basin. MB represents all elements of water resource modelling: users, reservoirs, hydropower, surface water, groundwater, rainfall-runoff and water quality[16]. By default, MB aims to study water allocation within a basin; however, a water quality option and a module for simulating groundwater can also be selected. The first step in setting up a river basin model is to define an underlying model river network upon which the node references can be based. Overlays with other features used in ArcView outside MB can also be performed, e.g. shaping files of river systems, thematic maps, etc. Water demand as processes/activities (irrigation area and public/industrial water supply) can be incorporated into the model. The model allows users to define the priorities of river diversions and water extractions (water rights) from multiple reservoir systems, as well as priorities for water allocation to multiple usages from individual extraction points. The model output comprises information on the performance of each individual reservoir and irrigation scheme within the entire simulation period, illustrating the magnitude and frequency of water shortages. The Groundwater module consists of a simple physical model of an aquifer, which is conceptualized as a linear reservoir that exchanges water with water users and surface water bodies. Water balance is analyzed according to pumping, recharge, seepage from rivers and discharge to rivers. The first three are assigned by the user as time series. Furthermore, time series of river flow at all nodes are simulated, enabling users to determine the combined impact of selected schemes

[14]AGWA is an ArcView extension which parameterises the SWAT and KINEROS hydrologic simulation models, runs the models and visually imports the results.

[15]DHI Water & Environment is an independent research and consultancy organisation, formed by a merger of the Danish Hydraulic Institute, VKI - Institute for the Water Environment and the Danish Toxicology Centre, see also http://www.dhigroup.com/Contact/AboutDHI.aspx.

[16]The MB illustration is based on tool manuals and the website http://www.dhisoftware.com/mikebasin/

on river flows. In order to tailor MB to the requirements of the WFD, attributes for representing the range of treatment options at all point sources are currently being incorporated, and the model itself has been extended by a water quality module. The soil erosion assessment module uses two well-known methods for simple source erosion assessments: the Universal Soil Loss Equation (USLE) and the Revised Universal Soil Loss Equation (RUSLE). Non-point pollution, including N and P-pollution from mainly agricultural areas, may be simulated in different ways at a more advanced level using the Daisy model.

RIBASIM 'River Basin Simulation Model' is a generic model package to analyze the behaviour of river basins under various hydrological conditions, produced by Delft Hydraulics. The software includes a graphical user interface, a database, a simulation program and a tool for the analysis of results. The model package is a comprehensive, flexible tool that links the hydrological water inputs at various locations to specific water users in the basin. The tool describes a basin in terms of water sources and uses. It performs simulations of water allocation along a certain time horizon. It can be useful to identify possible water use conflicts between different types of users, such as farmers or industries, to study the sustainable development of the river basin itself and to plan adequate measures to solve conflicts or generally improve the status of the water resource.

WaterWare, 'Water Resources Management Information System', is an integrated, model-based information and decision support system for water resources management. WaterWare Release 5.1 is fully accessible from the internet. This DSS can address a wide range of issues, such as: determining the limits of development; evaluating the impact of new environmental legislation; deciding what, where and when new resources should be developed; assessing the environmental impact of water-related development; formulating strategies for river and groundwater pollution-control schemes, etc. The tool supports the integration of databases, GIS, simulation models, optimization models and analytical tools into a common, easy-to-use framework.

SWIM, the 'Soil and Water Integrated Model', is a simulation tool for hydrological cycles, erosion, vegetation growth and nutrient transport in watershed at the meso scale, developed in Germany. The aim of SWIM is to analyze climate change and the impact of land use change on hydrology and water quality at the regional scale. The tool is based on two previously developed models, SWAT (Arnold et al., 1993) and MATSALU, which is a system of four coupled models developed for the Matsalu Bay watershed in Estonia (Krysanova et al., 1989). SWIM includes modules from both predecessors, endeavouring to combine their advantages (hydrological submodel and GRASS interface from SWAT; spatial disaggregation scheme and nutrient modules from MATSALU), and to avoid over-parameterization. A simplified EPIC approach (Williams et al., 1984) is used to simulate all crops and natural vegetation, using parameter values for each plant type from the database.

SPAW, 'Soil-Plant-Air-Water', simulates the daily hydrology of agricultural fields and ponds, including wetlands, lagoons and reservoirs. The objective of the SPAW model is to understand and predict agricultural hydrology and its interaction with soils and crop production, without the undue burden of computation time and input details. Field hydrology is represented by daily climatic descriptions of rainfall, temperature and evaporation; a layered soil profile with automated water characteristics; annual crop growth; and management with crop rotation and irrigation. Pond, lagoon and wetland simulations that have agricultural watershed fields or producer operations as their water source provide daily inundation levels, controlled by multiple input and depletion processes. Data input and file selection are by graphical screens. Simulation results are both tabular and graphical. Typical applications include analyses of crop water status, deep seepage, wetland inundation duration and frequency, lagoon designs and water supply reservoir reliability.

SWIM, 'Soil Water Infiltration and Movement', is a tool for simulating the infiltration, evapotranspiration and redistribution of water, developed by CSIRO Land and Water in Australia. The overall purpose of the model is to address issues relating to the soil water and solute balance. As such, it is a research tool that can be integrated into laboratory and field studies concerned with soil water and solute transport. It is also eminently suitable for management and education.

In the Netherlands, the RIZA (Rijksinstituut voor Integraal Zoetwaterbeheer en Afvalwaterbehandeling) has developed a set of coherent models to support the policy management of groundwater resources in the country. These models compute the hydrological effects of interventions on saturated and unsaturated zones, as well as the effects on agriculture, drinking water supply and nature.

LEACHM, the 'Leaching Estimation and Chemistry Model', is a suite of simulation models describing the water and chemical regime in the soil root zone, developed and applied in Australia. The suite consists of four simulation models and several utilities. The simulation models utilize numerical solution schemes to simulate vertical water and chemical movement. They differ in their description of chemical equilibrium, transformation and degradation pathways. Water regimes, pesticides, nitrogen and phosphorus and salinity in calcareous soils are simulated.

Ecohydrological models are highly relevant to studying the impact of agricultural on water systems. For instance, they can be used to quantify diffuse losses of N and P. The EU-funded project EUROHARP compared nine different methodologies and created a total of 17 study catchments across gradients in European climate, soils, topography, hydrology and land use. These methodologies are applicable at the catchment scale and are currently used by European research institutes to inform policy-makers at national and international levels.

'Low Flows 2000' is a catchment-based water resource decision support tool for the United Kingdom. It is a decision support tool designed to estimate river flows at ungauged sites and to support the development of catchment and regional water resources. It is the standard software system used by the Environment Agency and the Scottish Environment Protection Agency to provide estimates of river flows, as represented by annual and monthly flow duration statistics, for any river reach in the UK.

NIRAMS, the 'Nitrogen Risk Assessment Model for Scotland', is a GIS-based model to calculate losses of nitrogen from diffuse pollution. Estimates of annual residual N (by crop) are leached to surface and groundwaters by hydrological flows and routed to the stream system. The system uses nationally available data sets on land use, soils, topography and meteorology. It was developed within the ArcView GIS. The model was delivered to the Scottish Executive Environment and Rural Affairs Department and the Scottish Environmental Protection Agency.

NL-CAT, 'Nutrient Losses at catchment scale', is a tool capable of simulating phosphorus and nitrogen losses in both the soil and surface water, developed and applied in the Netherlands to replace the previous model STONE (Wolf et al., 2003). The model chain was constructed to evaluate the impact of different types of measures on the improvement of the surface water quality at the catchment scale. The surface water model consists of the two important key components, the 'Soil Water Atmosphere Plant' (SWAP) and the 'Agricultural NItrogen Model' (ANIMO), in combination with a surface water quantity and quality module. The SWAP module generates hydrological input to the ANIMO module, and simulates the nutrient cycle in soil and the nutrient leaching to groundwater and surface waters. Surface water discharges are simulated by a specific module, while simulation of surface water quality processes and retention within a (large) catchment is performed by the 'nutrient cycles in small surface waters' (NUSWALITE) module. An erosion module based on the modified and revised Universal Soil Loss Equations is implemented in a GIS environment. NL-CAT is capable of simulating phosphorus and nitrogen losses in both the soil and surface water.

DRIPS, 'Drainage runoff input of pesticides in surface water', is a GIS-based DSS, developed on behalf of the German EPA (Environmental Protection Agency/ Umweltbundesamt) for the exposure assessment of agriculturally used pesticides in surface waters. The tool estimates the quantity of pesticide input from non-point sources via surface runoff, tile drainage and spraydrift. Furthermore, the resulting predicted environmental concentration of pesticides in surface waters

(PECsw) can be calculated, considering the mean daily inputs of substances into river basins, characterized by their daily discharge. A graphical user interface (GUI) was created to provide users of the DSS with easy access to the model's algorithms. Model parameters can be modified by users in order to generate customized scenarios, predicting PECs for a choice of field crops, orchards or vineyards. Results are available as grid cell maps for the territory of Germany. The model aims to enable users to predict expected pesticide concentrations in river basins, thereby creating estimates of the probability of a quality target being exceeded. The results, illustrated in maps, can be used to identify hot spots of diffuse pesticide input. The water management industry can base its own measurements on these findings (Röpke et al., 2004).

TRK is a tool developed in Sweden for nitrogen (N) and phosphorus (P) gross and net load calculations, retention and source apportionment. The TRK system supports calculations of concentration and area losses of diffuse sources (for N from arable land using the dynamic soil profile model SOILNDB), calculations of the water balance (using the distributed dynamic HBV[17] model) and N transport and retention processes in water (using the HBV-N model). The results are presented in the GIS, and source apportionment is carried out for each sub-basin as well as for whole river basins. Results from the system have been used for international reports on transport to the sea, the assessment of the reduction of the anthropogenic load on the sea and for guidance on effective measures to reduce the load on the sea at the national scale. The tool is applied in Nordic countries and Sweden.

WetSpa, 'Water and Energy Transfer between Soil, Plants and Atmosphere', is a GIS-based hydrologic model that simulates hydrological processes continuously, both in space and time. Its aim is to simulate the hydrological behaviour of catchments with regard to flood prediction, land use and climate change scenario analysis and water management. The pollutant transport module aims to simulate the transport of phosphorus through a catchment. Domains covered include hydrology, land use, climate change, water management and phosphorus transport.

Other commercial tools for modelling water resources are distributed by:

- EMS-I 'Environmental Modeling System – Inc', including the 'Groundwater Modeling System' (GMS software), 'Surface-water Modeling System' (SMS software), and 'Watershed Modeling System' (WMS software). For further information, see: http://www.ems-i.com/index.html.

[17]HBV is the abbreviation of the (Swedish) name of the department of Sweden's Meteorological and Hydrological Institute, where the model was developed.

- BOSS International offers hydrology and hydraulics software - http://www.bossintl.com/.

5.3 Land use and economic tools

Another important group is represented by spatial models, where integration is again the paradigm. Integration of economic and ecological information in a spatial context is recognized as a valuable approach for strategic policy development and decision-making (e.g. Tiwari et al., 1999). Several conceptual frameworks encompassing the science, methods of capturing data and responses, and the human-biophysical dimension of problems have recently become the focal point of interdisciplinary research and analysis.

Lambin et al. (2000) categorize land use (LU) and land cover change (LC) models into five classes:

1. Empirical-statistical: uses multivariate statistical analyses of relevant factors to identify the causes/drivers of LU change. This type of model suffers from obvious limitations, since there is no reason why any statistically identified set of LC factors will apply robustly (or at all) outside the region or space/time slice from which they were identified.
2. Stochastic: a transition matrix quantifying probability describes how different LU types change from one to another. Since probabilities are determined from an analysis of the past, they can suffer from the empirical problems of a lack of robustness/generality, and of modelling data rather than the world.
3. Optimization: LU change is modelled as a process, whereby one measure is optimized (e.g. income per unit land). The approach assumes that actors behave rationally and are in search of the best solution.
4. Dynamic process-based: a more complex type of model that uses representations of social, economic and biophysical processes to simulate LC. Such complexity can require substantial effort to develop, verify and validate the data. However, the models can overcome the problems of previous approaches, and provide more insight.
5. Integrated: this type of model combines the characteristics of the above four categories.

In recent years, there has been a rapid expansion of interest and research into spatial decision support systems (SDSS), to which all the previous methods are applied. To varying degrees, these approaches attempt to

- capture the system dynamics;
- deliver outputs as spatial data that define biophysical, economic and social constraints;

- use new methods to translate factor layers into standardized inputs for problem criteria definition;
- use new methods to capture uncertainty in the ranking of alternatives.

The consideration of socioeconomic decisions and impact components requires the inclusion of decision models that need to represent the key land use, water use and management decisions made in the catchments, e.g. agricultural production decisions, industrial and urban water use decisions, reforestation and urbanization decisions. "*The specific decisions to be simulated and the types of models used to represent these decisions will depend on the spatial and temporal scales at which these decisions are to be modelled as well as on the types of activities present in the catchments. For example, even where extractive uses such as irrigation direct from the stream are considered, this decision may be modelled differently depending on whether the decision is posed as a short-run decision, considering capital to be constrained, or a long-run decision where capital investment decisions are included in the model. Additionally for some issues a representative farm model, simulating decisions by an individual farm, may be used, whereas for larger scale studies, or studies where trade-offs between different industry users are to be considered, aggregated regional production models may be used. In either case, it is the relevant land and water use decisions that are being represented.*" Letcher et al. (2007).

The previous passage identifies a key element in the spatial scale to choose the correct economic model. In fact, such models follow the separation of the theory into two branches: "macro" and "micro".

Macroeconomics is the study of aggregated variables addressing the state of the whole economy. The focus is on price levels, employment levels, economic output in real and monetary terms, the quantity of money in the economy, overall consumption, savings, investment, wage levels, etc. The aim is to provide an insight into guiding the formation of economic policy. The following tools implement macro-economic theory in operational models:

- AGLINK is a multi-country and -commodity dynamic model of world agriculture, developed by the OECD Secretariat in close co-operation with its Member States. The overall design of the model focuses primarily on the potential medium-term influence of agricultural policy on agricultural markets. AGLINK is a partial equilibrium model, primarily of major OECD commodity markets. AGLINK estimates supply, demand and prices. Non-agricultural sectors are not modelled, and are treated as exogenous to the model. Since 2004, this modelling system has been enhanced by the FAO's development of the 'COmmodity SImulation Model' (COSIMO), which represents agricultural sectors in a large number of developing countries.

The AGLINK-COSIMO modelling system is currently one of the most comprehensive partial equilibrium models for global agriculture. The model is one of the tools used in the generation of baseline projections underlying the OECD-FAO Agricultural Outlook.

- CAPRI – the 'Common Agricultural Policy Regional Impact Analysis' is an agricultural sector model covering both the whole of EU27 and Norway at the regional level (250 regions) and global agricultural markets. The following environmental indicators are covered by CAPRI: balances for N, P and K; emissions of ammonia, methane and N2O; global warming potentials. CAPRI users work in research institutions and EU Commission services.
- WATSIM – the 'World Agricultural Trade Simulation Model' is a recursive-dynamic, spatial world trade model for agricultural commodities. In its current version, it covers 12 regions and 29 commodities. Simulations run from the year 2000 to 2010. Policies covered are ad-valorem and specific tariffs, tariff rate quotas (TRQs), safeguards (flexible tariffs), export subsidies and production quotas. These policies are explicitly modelled by formulating the model as an MCP (mixed complementarity problem). The most important application of WATSIM is the medium-term analysis of trade policy changes.
- GTAP – The 'Global Trade Analysis Project' is one of the most relevant projects in this field. GTAP is a global network of researchers and policy-makers who are conducting a quantitative analysis of international policy issues with the goal of improving the quality of the quantitative analysis of global economic issues within an economy-wide framework.

Micro-economics explores economic agents, in this case farmers, and their behaviour and interrelation. The interpretation of a system's behaviour is an important classification criteria:

- Neoclassical economic theory describes decision-makers (e.g. farmers) as income maximizers, which implies a homogeneity assumption in preferences (Varela-Ortega et al., 1998; Iglesias et al., 2004).
- In an alternative multicriteria (MC) perspective, the farmers' decision-making processes are simultaneously driven by various attributes related to economic, social, cultural and natural dimensions (Amador et. al, 1998; Romero, 1989). Such attributes include the maximization of income, the minimization of risk, the maximization of leisure time, the minimization of managerial problems, etc. (Sumpsi et al. 1996). A decision-maker tries to satisfy all these criteria at the same time, and the choice is a compromise

reached despite conflicting objectives (Berbel and Rodriguez, 1998; Gomez-Limón et al., 2002).

- Other approaches follow a system's behaviour paradigm, including: agent-based modelling, cellular automata and related models that focus on complex spatial interactions; network theory and dynamical systems.

The previous approaches are adopted by different tools working at the micro scale, where a farm model describes the production possibilities (e.g. agricultural practices and management options) by engineering production functions that follow an approach, known as the primal representation of technology. With this approach, inputs and outputs including externalities are expressed in physical terms. This implies the possibility to measure and evaluate the effects of different sources of change on the environment as well as on the socio-economic dimension. The approach allows changes in soil and irrigation management practices, changes in the use of farm input and natural resources, and how agro-environmental measures and environmental regulations may constrain actions taken by farmers (Yaron and Dinar, 1982; Flichman et al., 1995; Doppler et al., 2002; Iglesias et al., 2003; Yang et al., 2003; Bazzani et al., 2005). An alternative approach is 'positive mathematical programming' (PMP), which derives unknown production functions from observed choices through the statistical calibration of the model (Howitt, 1995). PMP has generated numerous applications and extensions at different investigation levels, several of which are reported in Heckelei and Britz (2005).

The EU WADI project ('The sustainability of European irrigated agriculture under Water Directive and Agenda 2000') is a recent example of economic modelling in agriculture to analyze the sustainability of irrigated agriculture in Europe in the context of post-Agenda 2000 CAP Reform and of the Water Framework Directive (website: http://www.uco.es/grupos/wadi/). The research, conducted in five European countries, adopts a MC model-based approach, focusing on representative farms. Scenarios are used to anticipate the possible future state, related to macro-economic and institutional conditions. Irrigated agriculture is assessed in terms of sustainability, taking into account the multi-functional dimension. A set of indicators offers an insight into economic, social and environmental dimensions. The agricultural model is composed of a collection of microeconomic mathematical programming models, each representing the optimizing farmer's behaviour at the farm level. Simulation results are aggregated according to farm localization, type and size at higher territorial levels. The conclusion of the WADI project deserves attention: *"Future demands for agricultural economists and operational researchers . . . imply a demand for models capable of generating detailed microanalyses based upon diversity in farming systems. Such diversity appears to be greater than ever, as conventional agriculture faces up to organic farming, and as various technologies (GMOs, etc.) compete for consumer markets; global equilibrium may not reflect diversity in the adaptation of strategies and the impact of measures. For this reason, we need to encourage the development and adoption of simple models of irrigated agriculture that are capable of simulating changing policy scenarios and measuring the impact of such changes on social, economic and environmental indicators. . . . Agricultural economic models may help us to understand the value and cost of water and the multiple links between uses, consumption and pollution, by simulating the way in which these aspects are connected through the farming systems and their adaptation to external scenarios . . . modelling can support the definition of institutional measures that will facilitate the achievement of objectives (e.g. markets, quotas, economic support), by identifying incentive- and institutionally compatible alternatives and supporting their choice through simulations . . . Models could play a major role in the analysis and management of conflict of water use as well as in the analysis of the potential benefits of cost-sharing."* Berbel and Gutierrez (2004) pg. 217.

A recent contribution analyses the implications and applications for the European WFD of hydro-economic modelling in river basin management (Heinz et al., 2007). Four main applications are discussed:

- Economic, social (e.g. employment) and environmental impacts of water pricing.
- Cost-effectiveness analysis of integrated management strategies for pollution control.
- River basin modelling to analyze resource and environmental costs.
- Participatory planning.

Many existing tools have added socio-economic modules to their environment, and economic studies have also ensued (see: Ward and Lynch, 1996; Rosegrant et al., 2000; Cai and Rosegrant, 2004; Roetter et al., 2007). Examples include:

- 'Decision Support for IRRigation' (DSIrr) is a support tool for scenario analysis using simulation approaches and bio-economic optimization models. Scenarios evaluate a full range of water development and management options, taking account of multiple and competing uses of water. Scenarios can also describe agricultural policy or other exogenous drivers. A multi-scale modelling approach is adopted, and different types of farms can be aggregated to create a model network. The individual models are specialized to address specific questions and reflect heterogeneity between farms. Farm preferences can be described using a multi-attribute approach. Changes in irrigation and farm practices are considered. Standard procedures assess the impact of alternative water pricing schemes and of reduction in water allotment. The tool allows the CAP to be simulated, including decoupling subsidies, eco-conditionality and agro-environmental measures. However, case-by-case implementation is required. A set of indicators is quantified to assess socio-economic and environmental impacts at farm and regional levels. The tool, which was developed and applied in Italy, can easily be linked to other models.

- FARMIS assesses the impact of the Mid-term Review policy reform on German agriculture by simulation approaches and optimization models based on the 'Farm Accountancy Data Network' (FADN)[18]. The model network includes different models that are updated and combined with each other as necessary. The individual models are specialized to address specific questions (for example, regional impacts, farm impacts, market impacts). Comparative-static analysis is conducted, where the main purpose is to analyze the sectoral effect of policy measures addressing the farm level.
- The 'Interactive Component Modelling System' (ICMS) is a framework in which to embed models and tools for the analysis and presentation of environmental options for environmental managers. It applies scenarios as model inputs, which take into account achievable policy/management drivers and socio-economic capacity for change, as well as uncontrollable system shocks. The DSS computes indicators, covering a spectrum of biophysical and socioeconomic impacts for each scenario as model outputs. The tool was developed in Australia and is being applied in Asia.
- The 'Integrated Water Resource Assessment and Management' (IWRAM) software contains two modelling toolboxes that utilize a nodal network structure for catchment analysis: a Biophysical Toolbox (erosion, streamflow, crop) and an Integrated Modelling Toolbox, which links models of household decision-making to allow for the consideration of socio-economic and environmental trade-offs of many development and policy scenarios. Previous applications were performed in Northern Thailand.
- The 'Spatial Decision Support System' (SDSS) provides watershed economic analysis by maximizing the profit of a representative farm assumed to cover the whole watershed with the constraints of production

[18]FADN consists of an annual survey carried out by the Member States of the European Union. The services responsible in the Union for the operation of the FADN collect accountancy data every year from a sample of agricultural holdings in the European Union. Derived from national surveys, the FADN is the only source of micro-economic data that is harmonised, i.e. the book-keeping principles are identical in all countries.

technology, resource, sediment control objectives and sustainable utilization. The tool includes two major types of models: static and dynamic. Each model type supports variations in plant growth, grazing and ranch operations. Upland erosion is estimated through RUSLE2 and the sediment yield of a watershed is estimated from upland erosion and sediment delivery ratios. The tool was developed and is being applied in the United States of America.

- 'WaterStrategyManDSS' (WSM DSS) is a GIS-based Decision Support System developed for the WaterStrategyMan Project. It aims to assess the state of a water resource system in terms of sources, usage, water cycles (pathways) and environmental quality. In addition, it evaluates the effects of actions on the basis of different scenarios, alternatives and policies. Water allocation is performed according to a set of demand and supply priorities, reflecting pricing systems, social preferences, environmental constraints and development priorities. WSM DSS includes the following types of management options: supply enhancement options, which are intended to increase available water quantities during drought, are concerned with structural interventions that attempt to enhance water supply; demand management options, aiming to decrease water demand through various conservation techniques and use limitations; socio-economic measures to mitigate impacts, such as pricing and changes in regional development; methods to produce management strategies through combinations of control measures seeking optimum and efficient solutions. Applications have been carried out in the Mediterranean region.

A tool that adopts a different approach is PlayAgriPoliS, a policy simulation game. The tool is an agent-based model that establishes a link between agricultural policy reform and structural changes. Although PlayAgriPoliS is currently under development, a beta version is currently available for download.

The 'Water Evaluation and Planning system' (WEAP) is a user-friendly software tool applied in the USA that follows an integrated approach to water resource planning. The tool supports the identification and evaluation of the impact of climate change on water for agriculture, and entails alternative strategies to study the cost of water in watersheds. The tool supports the assessment of water supply augmentation through an inter-basin transfer within a firm yield analysis and the development of supply and demand balances.

To support policy, the socio-economic sub-system needs to include a component that describes the relevant social and economic impact of change in other system components, such as the impact on farm income and employment. In some cases, local impacts are aggregated and transferred to a higher scale model to assess the impact on the regional economy. Again, the scale and range of impacts to be considered dictates the type of modelling approach used.

The choice (and subsequent implementation of the chosen option) represents the final phase of the decision-making process. This stage can be supported by multi-criteria decision analysis (MCDA). Once the criteria for evaluation have been determined, options can be compared and assessed against their expected impacts using Multi-Criteria Analysis (MCA) evaluation techniques.

The European school of MCDA (Roy, 1996) has created extensive literature and diverse methodologies for the application of MCDA, including the ELECTRE[19] family. Alternative approaches to MCDA are the Analytical Hierarchy Process (AHP) (Saaty, 1980) on the one hand, and an array of cognitive methods on the other. These approaches address the decision-making process in detail, and deal with a limited, clearly defined set of alternatives[20].

Many computer-based approaches have been developed to deliver MCDA, or elements thereof, in a range of forms, e.g. MACBETH[21], Intelligent Decision System IDS[22] routines in IDRISI[23] or MULINO-DSS, and more recently in NetSyMod (Network Analysis - Creative System Modelling - Decision Support), which is an operational tool that aims to support and guide decision-makers at each step of the overall decision-making process, from problem conceptualization to the choice of the best policy to solve it.

LADSS, the 'Land Allocation Decision Support System', is a farm-scale land use planning tool being developed in the United Kingdom to assist in the case-based investigation of policy and environmental change impacts on land use systems. The tool supports strategic, farm-scale, land use planning by suggesting possible combinations of land use to meet multiple objectives. LADSS provides a framework within which the financial, social and environmental consequences of changes in land use can be evaluated. It acts as a channel for technology transfer from land use scientists to land managers, and facilitates the inclusion of practitioner knowledge into models of land use systems.

The 'DSS for the Elbe River Water Quality Management' includes models, spatial and non-spatial data and analysis tools under a user-friendly GIS-based interface. The tool confronts decision-makers with possible measures, as well as multiple management objectives. DSS helps water managers to formulate policy for river

[119]ELECTRE 'Classement et choix en présence de points de vue multiples'.

[20]Optimisation approaches are another important branch of the MCDA, which in some ways follows a bounded rational paradigm (Spronk and Veeneklaas, 1983).

[21]See: http://www.m-macbeth.com/Msite.html

[22]See: http://www.e-ids.co.uk/

[23]For further information see http://www.clarklabs.org/.

basin management and to take appropriate action to realize policy objectives. Furthermore, the DSS is especially suited to support participative decision-making.

Another experience is represented by the management of 'Regional German River Catchments' (REGFLUD), a project that addresses the problem of diffuse pollution. The overall objective is to develop and apply multi-criteria scientific methods to set up a DSS aimed at reducing diffuse pollution in river catchments, subject to economic feasibility and social acceptability. A model network is constructed, consisting of an agricultural sector model, a water balance model and a denitrification model that enables policy measure analyses to be performed. Alternative agro-environmental measures were carried out by a benefit-cost approach based on interviews regarding the social acceptability of alternative measures. Alternative forms of farm management regarding nutrient surplus were analyzed as a guide for water management.

AQUATOOL is a generalized DSS for water resource planning and operational management, developed in Spain. The model simulates the operational management of the system on a monthly basis. It is responsible for water allocation to water uses, and considers the connected use of surface water and groundwater. Operating policies are defined by the following variables: target, minimum and maximum volumes of reservoirs, inter-reservoir relationships and priorities of use, minimum flow in rivers, flow requirements for hydroelectric plants, targeted water demand for each agricultural, industrial and domestic area and demand priorities used in water allocation.

AgriBMPWater is a tool designed to compare Best Management Practices (BMPs) in terms of environmental efficiency, economic cost and potential acceptability by farmers. It adopts negotiation-based implementation methods.

DANUBIA is a DSS currently under development in the 'Global Change in the Hydrological Cycle' (GLOWA)[24] - Danube project. Upon completion, DANUBIA will be able to simulate water-related issues of environmental management under ecological, economic and cultural aspects, such as flood risk and protection, agriculture, water quality and quantity, tourism and water, as well as water and climate. It will examine the sustainability of the proposed solution scenarios.

SIGRIA, the 'Information System on Water Management for Irrigation', is an integrated set of tools developed in Italy by the 'Istituto Nazionale di Economia Agraria' (INEA) for the Ministry of Agriculture and for regional governments. SIGRIA is used: to plan new infrastructure, including dams and water distribution

[24]The German Federal Ministry of Education and Research (BMBF) launched the GLOWA programme to identify new research strategies in the field of global change, in particular with regard to the aim of sustainable development.

networks; identify and solve competitive situations for water by different users; mitigate the effects of annual droughts; develop and implement new policies on water pricing. At the local level, SIGRIA is a key component of information systems for irrigation managed by reclamation consortia. SIGRIA is based on GIS technologies and includes: a detailed map on land cover/use, identifying all irrigated areas and crops grown per season; an irrigation suitability map; all irrigation water networks, from the water source (dam river, etc.) to the farm or group of farms, including a database on the technical features of the network; extensive databases on the features of irrigation (i.e. crops, farms, costs) at the local level; a linear programming model on the optimum allocation of resources for irrigation for certain areas, at farm and basin levels, and a model on local water requirements by crops, depending on soil and weather data.

5.4 Web portal on water-related tools

Water managers, decision-makers, extension agents, regulatory agencies, planning organizations, consultants, students, researchers and environmental groups pose a number of questions, such as:

- What new models are available that I may not be aware of?
- What modifications and new versions have been added to my favourite model?
- Are new user interfaces, general databases or other time-saving devices available?
- Which applications can I take as a reference to address my problem in water management or water policy?

Consistent and comprehensive answers to the previous questions are essential for the practical and wide-spread application of models and tools.

Many existing models have been used or developed in the context of projects funded by the EU in recent years, and others are soon forthcoming (see Annex A for a list of relevant projects).

The importance of spreading existing information to all actors, updating it and making it easily available is crucial. For this reason, the EU has funded specific projects to create a European-wide comprehensive, shared web-based data and information management system for water, including river basins, with the primary purpose of consolidating information for potential users (Table 4).

Other water-related web resources are:

- 'The Register of Ecological Models' (REM), a meta-database for existing mathematical models in ecology. It is a co-operative service of the University of Kassel and the GSF - National Research Center for Environment and Health) - http://eco.wiz.uni-kassel.de/ecobas.html

Table 4 European web portal, including toolbox for water

Tool name	Water Information System for Europe	Harmonised Modelling Tools for Integrated Basin Management	Harmonising Quality Assurance in model-based catchment and river basin management	River Basin Manager's Toolbox - Model Evaluation Tool
Acronym	WISE	Harmoni-CA	Harmoni-QuA	MET
Description	WISE is a portal that compiles data and information collected at EU level by various institutions or bodies.	Harmoni-CA is a European-wide forum bringing together the scientific and political world involved in integrated basin management and the implementation of the WFD	The HarmoniQuA project has developed a computer-based Modelling Support Tool (MoST) to provide a user-friendly guidance and quality assurance framework.	MET aims to test and demonstrate the use of integrated models applied to selected intensively studied river basins.
Website	http://www.water.europa.eu/	http://www.harmoni-ca.info/	http://www.harmoni-ca.info/	http://www.rbm-toolbox.net/toolbox3/
Objective	The primary objective is to create a comprehensive, shared European dwtata and information management system for water, including river basins, following a participatory approach.	To establish a communication forum and define general methodology and guidance documents. Joint use of monitoring and modelling. Integrated assessment and the science-policy interface.	HarmoniQuA aims to provide a user-friendly computer-based guidance and QA framework for use in model-based river management.	The objective of the BMW project is to establish a set of criteria to assess the appropriateness of integrated models for use in the implementation of the WFD.

Additional information	WISE is a partnership between the European Commission (DG Environment, Joint Research Centre and Eurostat) and the European Environment Agency. The web portal http://www.wise-rtd.info/ links websites with a focus on information relevant to the implementation of the WFD, such as (CIS) guidance documents, selections of ICT tools, methodologies and results of research projects (e.g. the CatchMod cluster). Information is offered at European, national and regional levels, as well as for river (sub-)basins.		MoST is a DS that prompts users with the appropriate 'next step' in the modelling process and provides an audit trail to check previous decisions. The approach targets management at catchment and river basin scales with the overall goal of improving the quality of modelling and therefore enhancing the confidence of all stakeholders in them. The tool is freely available at: http://harmoniqua.wau.nl/public/Product/software.htm	The MET toolbox provides information on models and other tools required in the implementation of the Water Framework Directive (WFD) and helps model users to select appropriate tools for their specific needs. The Model Catalogue provides information on the characteristics of various models, specifically designed for questions and problems arising in the context of the WFD.

Table 5 European web portal, including toolbox for water

Tool name	EUROHARP	Standardisation of River Classifications
Acronym	EUROHARP	STAR
Description	The Toolbox provides factual and searchable information on the nine quantification tools (models) and the 17 European catchments tested within the EUROHARP project.	Framework method for calibrating different biological survey results against ecological quality classifications to be developed for the Water Framework Directive
Website	http://euroharp.org/index.htm	http://www.eu-star.at/
Objective	The primary objective is to provide end users (national and international European environmental policy-makers) with a thorough scientific evaluation of nine contemporary Quantification Tools for diffuse losses.	Supporting multidisciplinary teams in projects that use models for water management
Additional information	Quantification Tools within this framework are methodologies for quantifying diffuse losses of nutrients (N and P) from land sources to surface waters. They range from simple empirical and statistical models to more advanced scientific process-based models. Different Quantification Tools have been developed in different countries. They vary significantly in their (i) level of complexity, (ii) representation of system processes and pathways, and (iii) resource requirements (data and time). Quantification Tools should aim to be accurate and responsive to changes in land use and land management. Furthermore, they should be influenced by geophysical factors (topography, hydrology), biochemical factors (soils) and climatic factors (rainfall, temperature), as well as providing end users (policy-makers, implementers and evaluators) with information at appropriate time scales to assess compliance with remediation measures.	The STAR software comprises programs for - data storage and the evaluation of hydromorphological features (STAR RHS) - data storage and the evaluation of macrophyte samples (STAR MTR) - data storage of invertebrate and diatom samples (AQEMdip) - metrics calculation and the data evaluation of invertebrate samples (AQEM European Stream Assessment Program) - uncertainty/error estimation for assessments of ecological status class based on multiple metrics and EQRs (STARBUGS) - guidance in devising monitoring programmes (MONSTAR)

- The Hydroarchive website is an open source toolbox for model optimisation, evaluation and hydrological modelling, including model codes for rainfall runoff modelling - http://www.sahra.arizona.edu/software/index_main.html.
- The Hydrologic Modeling Inventory is maintained through a cooperative effort between the Bureau of Reclamation and Texas A&M University- http://hydrologicmodels.tamu.edu/
- The West National Technology Support Center (WNTSC) for the USDA Natural Resources Conservation Service (NRCS) - http://www.wsi.nrcs.usda.gov/products/W2Q/H&H/Tools_Models/tool_mod.html
- The Archives of Models and Modeling Tools[25] of the 'Surface-water quality and flow Modeling Interest Group' (SMIG) of the US Department of the Interior | US Geological Survey - http://smig.usgs.gov/SMIG/model_archives.html
- The Surface Water and Water Quality Models Information Clearinghouse (SMIC) is a database of the descriptions and features of environmental surface water and water quality models, and abstracts of projects using such models http://smig.usgs.gov/SMIC/

A quite different tool is AQUASTAT – the FAO's global information system on water and agriculture. This information system consists of many databases, including: dams, institutions, river sediment yields, the state of water resources and agricultural water use by country and region, climate, a global map of irrigated areas http://www.fao.org/nr/water/aquastat/main/index.stm

As far as water and agricultural models and tools are concerned, two new projects are relevant:

- The first, 'Open Modelling Interface and Environment' (OpenMI), created a standard for model linkage in the water domain. It was developed by the EU-funded HarmonIT project[26]. Specific objectives are that the mechanism's design should:
  - be applicable to new and existing models;
  - impose as few restrictions as possible on the modeller's freedom;
  - be applicable to most, if not all, time-based simulation techniques;
  - require a minimum of change to the program code of existing applications;
  - keep the costs, skills and time required to migrate an existing model to a minimum;

[25]It would appear that the websites have not been updated.

[26]The project was funded within the Fifth Framework Programme and contributes to the implementation of the Key Action "Sustainable Management and Quality of Water" website: http://www.openmi.org/

- be easy to use;
- not unreasonably degrade performance.

The assumption is that integrated catchment management requires integrated analysis that can be supported more effectively by integrated modelling systems. These can be developed and maintained better if they are based on a collection of interlinked models. OpenMI intends to launch "a new era in integrated water management", as quoted from their main website.

- The second, the 'System for Environmental and Agricultural Modelling; Linking European Science and Society' SEAMLESS, is an EU FP6 Integrated Project set up in response to a research and policy need formulated by the European Commission. The project aims to develop an integrated framework called SEAMLESS-IF that allows the ex-ante assessment of agricultural and environmental policies and technological innovations. The framework will have multi-scale capabilities, ranging from field and farm scales to the EU25 and the globe; it will be generic, modular and open, using state-of-the art software[27]. This computer system will include models that simulate effects on the environment and on economic developments, besides procedures that enable the assessment of the social aspects of sustainability. SEAMLESS-IF will assess and compare, ex-ante, alternative agricultural and environmental policy options, allowing:
  - analysis at the full range of scales (local farm scale to EU and world scales), whilst focusing on the most important issues emerging at each scale;
  - analysis of the environmental, economic and social contributions of multifunctional agriculture to sustainable rural development and rural viability;
  - analysis of a wide range of issues and agents of change, such as climate change, environmental policies, rural development options, effects of an enlarging EU, international competition and effects on developing countries.

The tool will enable the translation of policy questions to alternative scenarios that can be assessed through a set of indicators capturing the key economic, environmental, social and institutional issues of the questions at stake. The indicators in turn are assessed using an intelligent linkage of quantitative models designed to simulate aspects of agricultural systems at specific scales, i.e. point or field scales, farm, region, EU and world scales. Application of the models requires pan-European databases for environmental, economic and social issues. Some indicators, particularly social and institutional ones, will be assessed

[27]The project is being carried out by a consortium of 30 partners, led by Wageningen University (NL), e-mail: seamless.office@wur.nl website: http://www.seamless-ip.org/

Table 6 Types of models used in SEAMLESS

Bio-physical models:
- Spatial scale: plot of homogeneous soil with a defined weather structure, including variability over time.
- Time scale: long-term, to take into account cumulative effects on natural resources and the environment. Recursive structure: the past influences the future, but the future does not influence the past, as in inter-temporal optimization models.
- Biophysical processes and human intervention are modelled interactively.
- In terms of information flows, the 'Agricultural Production and Externalities Simulator' (APES) receives data from nature (i.e. climate, soils, etc.) and from human activities (weather, soil, techniques, rotations, etc.) and provides information on a very wide range of outputs (yields, pollution, change in the soil state, etc.).
Economic models (bio-economic farm models – 'Farm System Simulator' FSSIM):
- Spatial scale: the farm, as a decision unit composed of several plots of soil allowing for heterogeneity.
- Time structure: can be static or dynamic. If dynamic, it can be recursive (independent optimization for each period, the past influences the future) or inter-temporal optimization (optimization of a flow for a defined time horizon) or dynamic-recursive (a mixture of these two).
- Positive approach: models should be able to reproduce real situations and react to exogenous and endogenous changes (policies, effects of agricultural practices on productivity of natural resources)
- Information flows: some information comes from APES: yields, emissions, impact on soil erosion, etc. Other information comes from statistical sources, surveys, users' requirements: prices, labour use, costs, size of farms (DATA BASES) and policy framework. At the farm level, bio-economic modelling, prices and policies are clearly exogenous. However, there are plans to use an iterative loop with the market model in order to define prices in a specific simulation exercise. On the other hand, links to bio-physical models may also require feedback.
Market models:
- Agricultural Sector Model (CAPRI), Global Model (GTAP).
- Spatial scale: European Union for CAPRI, the World for GTAP. There is also a world market module in CAPRI.
- Time structure: comparative-static.

Table 6 *Continued*

- Positive approach.
- Information flows: some information comes from FSSIM for estimating the supply side of CAPRI. Other information comes from statistical sources, surveys, users' requirements, prices, labour use, costs, size of farms (DATA BASES) and policy framework. (At the level of market modelling, prices are endogenous.)
Landscape models:
- Spatial scale: regional.
- Time structure: static.
- Information flows: in principle, these models will be "passive". They should be able to reflect the impact on landscape produced by the results obtained from simulations performed using the other models.

directly from data or via a post-model analysis. The software backbone of the SeamFrame project will link models designed for different scales and domains, and will facilitate the reuse, maintenance and documentation of models. The types of models[28] used in SEAMLESS are shown in Table 6.

Prime users of the SEAMLESS Integrated Framework are the Directorates General (DG Agriculture, Environment and Economics and Finances) of the European Commission, the JRC and the European Environment Agency (EEA). It is also the intention to make the tool useful to national/regional policy-making agencies, farmers' organizations (e.g. COPA) and NGOs. A third important future user group is the scientific community.

6. CASE STUDIES: EXAMPLES OF GOOD APPLICATION

Key message:

Three examples of good application are presented. The following selected case studies have all been recently implemented in the context of EU-funded projects in different European water basins:

- WaterStrategyMan in the ARID cluster
- The German Elbe River Basin
- The Piave River Basin in Italy

[28] A description of the types of models used in SEAMLESS and selected characteristics can be found in SEAMLESS No. 010036 Deliverable number: PD1.2.1, 2 June 2005.

> The case studies offer an insight into relevant questions such as: How can tools improve transparency in decision–making? How can tools support water managers in their present tasks? How can different management options be compared in terms of their ecological, ecnomic and social impact.

This section contains three reference case studies on how tools can support the implementation of the principles of the WFD in agriculture.

The selected case studies have all been implemented recently in the context of EU-funded projects in different European water basins.

6.1 ARID cluster and WaterStrategyMan

The so-called ARID cluster[29] includes three research projects on integrated and sustainable Water Resources Management in arid and semi-arid regions:

- WaterStrategyMan (WSM - 'Developing Strategies for Regulating and Managing Water Resources and Demand in Water Deficient Regions'), website: http://environ.chemeng.ntua.gr/wsm/
- Medis ('Towards Sustainable Water Use on Mediterranean Islands: Addressing Conflicting Demands and Varying Hydrological, Social and Economic Conditions'), website: http://www.uni-muenster.de/Umweltforschung/medis/
- Aquadapt ('Strategic Tools to Support Adaptive, Integrated Water Resources Management under Changing Utilization Conditions at Catchment Level: A Coevolutionary Approach').

The first project was selected because it represents a good example of how tools can support participatory processes to develop strategies for regulating and managing water resources and demand in water-deficient regions. The project recognizes that the existing framework of water management infrastructures, the natural environment, water supply (production) and consumption, institutional and socio-economic conditions are the main factors in determining the appropriate strategies that may lead to improved water resources management. Interactions between the main components of the socio-economic and environmental systems determine the issues analyzed by the WaterStrategyMan DSS. With the assistance of GIS and properly customized databases, the objective of the DSS is to:

---

[29]Arid cluster comprises three projects supported by the European Commission under the Fifth Framework Programme, contributing to the implementation of the Key Action "Sustainable Management and Quality of Water" within Energy, Environment and Sustainable Development. The main source for this example is represented by the respective websites.

- describe the existing situation in a case study area, in terms of the hydraulic and environmental characteristics of manufactured and natural water systems;
- assess the state of a water system, addressing different aspects, such as available water sources, actual usage practices, water cycles (paths), environmental quality and economic issues;
- define and apply alternative strategies for integrated water resources management, built on technical management options/actions;
- forecast the behaviour of the state of the water system, on the basis of assumed or envisaged scenarios of water availability, and water demands for different types of uses;
- evaluate the impact of actions by observing and analyzing the results of forecasted scenarios, alternative options and policies, through a multi-criteria evaluation approach and consideration of local, national or international legal constraints.

The DSS considers the differences between quantity and quality dimensions in water management and the development of alternative options and long-term scenarios. It establishes a broad framework on the existing knowledge on IWRM practices, while highlighting the importance of regionalization and the relevant cultural context. The economic analysis in the WSM DSS (Water Management Schemes DSS) consists of a tentative implementation of the principles associated with the estimation of Full Water Costs and its components. The aim of the Evaluation Module is to facilitate the comparison of alternative WMSs, which incorporate different scenarios and/or strategic options. Each scheme's score is defined as the value of a Relative Sustainability Index, which is derived from specific statistical criteria that are calculated for each scheme and based on a predefined list of indicators.

The project has made a great contribution to the development of methodologies to support the economic analysis requirement of the WFD. It pays special attention to the evaluation of direct and indirect costs of water use, the estimation of appropriate water prices and the comparison of alternative water management scenarios, using an integrated multi-criteria approach that is able to represent general guidelines for improved water management practices in water-deficient regions.

6.2 The German Elbe River Basin case study[30]

The German Elbe river basin project was also a success. River basin management has become increasingly complex in recent decades. Demands made by society regarding the ecological and chemical quality of stretches of water, the use and

[30]The case study reflects the information given on the project's website.

protection of water bodies, pollution by numerous substances as well as flood protection issues have led to the development of new views and strategies with regard to (the making of) policy for river basin management. The objective of this project was to develop a DSS for the Elbe river basin. Such a DSS helps water managers to formulate policy for river basin management and to take appropriate measures to realize policy objectives[31]. Simulation models are used to assess the efficiency of measures, such as reforestation, changes to agricultural practices or the efficiency of wastewater treatment plants to achieve management targets. MONERIS ('MOdelling Nutrient Emissions in RIver Systems') and GREAT-ER ('Geography-referenced Regional Exposure Assessment Tool for European Rivers') are integrated into the Elbe-DSS to assess nutrient and pollutant loads. MONERIS calculates nutrient inputs from diffuse and point sources at a sub-catchment scale of about 1000 km^2. GREAT-ER is a tool to assess the exposure of point source emissions that considers fate in sewage treatment plants as well as degradation and transport in rivers. Both models make long-term predictions, but differ in their spatial scales. An ad hoc procedure has been created to link them. The DSS allows the distribution of diffuse nutrient emissions calculated from MONERIS and point source emissions from GREAT-ER to the river network, where further elimination and transport processes are calculated. The tool consists of a GIS-based user interface, which allows flexible easy-to-use access to predefined scenarios, measures and models. A database management system (DBMS), a model-based management system (MBMS) and a knowledge-based toolbox are integrated into the graphical user interface. Various data sources are integrated into the DSS, including: land cover (EEA, 2004), a digital terrain model (GLOBE Project), soil properties, hydrological and meteorological data (Federal Agencies), census data (Federal Statistical Office), wastewater treatment data and discharge consents (Federal and State Environmental Agencies) and monitoring data (various sources). All spatial data sets are processed with a GIS to produce a consistent geo-referenced data basis, which is coupled to the simulation models (Matthies et al., 2006).

A participatory approach involving local actors, experts and decision-makers was carried out to identify the issues to be addressed in the DSS:

- how to improve the socio-economic conditions of the different uses of the river basin and catchment (shipping, tourism, fisheries, agriculture, etc.);
- how to provide an acceptable and sustainable flood level;

[31]The study was carried out by the Universities of Twente/Enschede and Osnabrück and the institutes RIKS and INFRAM from the Netherlands, based on a feasibility study by the Federal Institute of Hydrology (BfG).

- how to achieve a sustainable improvement of the physical, chemical and biological state of the Elbe and its tributaries;
- how to reduce nutrients from agriculture, settlements and industries, and the chemical load into the North Sea;
- how to increase the ecological value of the river and floodplains in the Elbe river basin.

Packages of measures to be taken were selected for assessment, such as:

1. High water management: dike shifting and other measures
   - dike shifting (space for the river)
   - reduction of buildings and other edifices in the foreland
   - provision of information on high water management
2. Water quality
   - reduction of point- and non-point-source pollution by improving agricultural practices (nutrients, pesticides)
   - reduction of pollution by hazardous substances
   - improvement/installation of waste-water treatment plants
3. Navigability
   - groyne modification
   - dredging
4. Reduction of riverbed erosion
   - adding material to the river bed/sand suppletion
5. Nature protection
   - zoning (nature reserves)
   - development of near-nature floodplains
6. Tourism
   - several measures.

Results from the study show that various governments, non-governmental organizations and institutions have been able to benefit from the DSS. The Elbe DSS could act as a prototype for other river catchments.

6.3 The case study of the Piave River Basin in Italy

This study was developed within the EU-MEDA project ISIIMM ('Institutional and Social Innovations in Irrigation Mediterranean Management'), which focuses on irrigation and water management. It addresses a wide range of related issues, such as relationships with the environment, management policies and the adaptation of local societies to local problems. The Italian case study area is the

Piave River Basin. In the basin, a pilot area represented by a small zone located to the northwest of Treviso was identified for its relevance to the topic. In this area, reorganization of the current water management system is in urgent need, due to conflicts regarding use of the same water bodies for different purposes within and outside the area. In particular, a programme of measures is being developed to substitute the existing irrigation systems and network based on open canals and furrow irrigation with an integrated pressurized dual system to distribute non-potable water of high quality for irrigation and other services, such as fire prevention, car washing, etc. With regard to irrigation, the methods would shift from current low-efficiency surface infiltration to more efficient methods, adopting sprinkler irrigators at the farm level.

Against this background, a study was conducted with the general aim of:

- understanding the socio-economic characteristics of the farms located in the pilot area and in particular the economic convenience of farmers to replace the current irrigation system;
- assessing future trends in agricultural income, employment and water demand under different scenarios, taking into account the CAP reform.

The methodology adopted is based on simulation techniques implemented via bio-economic models in the 'Decision Support for IRRigation' DSIrr (Bazzani, 2005a, 2005b), integrated within the NetSyMoD approach to manage the involvement of stakeholders, build conceptual models, and decision support through the mDSS4[32] software tool.

Combination of the DSIrr and mDSS tools within the NetSyMoD allowed users to:

- develop and share alternative scenarios of potential local futures with stakeholders;
- involve stakeholders in the evaluation of the feasibility of reorganizing the integrated water distribution system;
- elicit and document concordant opinions on the proposed plan of measures;
- integrate stakeholder visions, opinions and preferences with bio-economic modelling into the same conceptual model of the local territorial system and its possible futures;
- identify the strengths and weaknesses of the proposed plan compared to the possible evolution of the area.

[32]In mDSS4, 'm' stands for multicriteria and the number 4 identifies the version.

The specific contributions of the two software tools were as follows:

- By integrating agronomic and engineering information with economic models and by favouring a joint consideration of water and agricultural policy, DSIrr enabled users to estimate ex-ante land use variations. This approach also allowed the consideration of equity issues concerned with the distribution of cost and benefit among farm typologies upon adoption of economic instruments, such as water pricing or other measures;
- mDSS allowed the evaluation of the two alternatives in question (whether or not to reorganize the distribution and irrigation systems) within the contexts of three possible scenarios, by integrating the criteria and weighting expressed by stakeholders at the NetSyMoD workshop and the quantitative socio-economic indicators, which were supplied by DSIrr.

The study showed that, even in a small region that is far from a homogeneous reality, agriculture is a complex system that can be described more accurately using a hierarchical decomposition approach. This complexity should be taken into account if the potential cost of water in agriculture is to be estimated and sound measures defined.

The adopted sustainability indicators highlight the side-effects of water policy in terms of their economic, social and environmental impact, and show that different farming systems have specific responses to water policy. In fact, the agricultural adaptation process integrates variation in land use, production processes, irrigation levels and technology, differentiated by type of farm. In addition, the scenario analysis clarifies the connection between water and agricultural policy.

The research confirmed the usefulness of the DSIrr simulation tool and its integration with mDSS into the NetSyMoD framework, to support decision-makers by integrating water and agricultural policy analysis. The conjoint analysis of the two policies proved to be essential to identify a sustainable path and design sound intervention for the agricultural sector, which considers cost recovery and the polluter pays principle and is sustainable to the agricultural sector, as well as being beneficial to the environment and society.

The case study can offer an insight into relevant issues, such as: how tools can improve transparency in decision-making, how tools can support water managers in their present tasks, how different management options can be compared in terms of their ecological, economic and social impact.

7. CONCLUSIONS

> Key message
>
> Models are widely used in modern river basin management throughout the world. Successful projects show that integrated tools require multidisciplinary teams of well-trained people and adequate data to be applied properly.
>
> The choice of tool is highly dependent on local circumstances, among which previous experiences in modelling, trained staff and data availability are critical.
>
> A multilevel approach in modelling, integrating common conceptual models for cross-disciplinary work involving authorities, managers, stakeholders and researchers, and quantitative models to carry out specific analysis exploring the sensitivity of the system to changes, is recommended.
>
> Good practice in model development and application deserves close attention. In fact, the credibility and impact of the information and insight that modelling aims to generate are highly dependent on the quality of the modelling exercise.
>
> Since innovations in computational technology will further enhance the frontier of modelling, dedicated web portals offering comprehensive, shared European information, such as WISE, are a key instrument for water and agricultural managers to access updated information.

The use of toolboxes and models is mentioned in several places in the Water Framework Directive document and in various guidance documents, although the use of models is not obligatory. A key issue is 'if' and 'how' tools can support water and agricultural managers in their effort to comply with WFD requirements. To answer the previous questions, this review describes over fifty tools related to water and agriculture, taking into consideration agronomic, hydraulic, land use and socio-economic dimensions. What emerges is a complex, unclear picture.

In the past decade, research has led to the creation of many tools, some of which have reached the stage of operational implementation and can address specific aspects of the complex relation between water and agriculture. In many cases, different tools address the same problems and issues, and the differences between models are minor.

The new computational technology continues to enhance the frontier of modelling. Most new tools are DSS, GIS-based, integrated and user-friendly, with a beautiful front end. Nonetheless, the embedded complexity is very high. There is a strong need, which goes beyond the scope of this review, to evaluate the conceptual core of such tools; the EU project EUROHARP is a good example of the specific field considered.

The WFD adopts a basin approach that can be supported by the last generation of integrated hydrological tools, which are primarily concerned with the representation of the physical world by GIS. Such tools are often weak with regard to economic analysis, which could be addressed using other models and tools.

The integrated modelling approach adopted by tools, including hydrology and water quality models, agronomic models, land use and economic models, requires expertise from various domains. They should be performed by teams of people from a variety of disciplines. This, however, could create new problems, as recent experiences have shown. A multidisciplinary approach could complicate working processes communication problems are often present and no common vision is guaranteed.

A growing need for quality assurance in modelling has emerged among professionals in this field (Refsgaard, 2005). Quality assurance is defined by NRC (1990) as the procedural and operational framework used by an organization to manage a modelling study to assure the technically and scientifically adequate execution of all tasks included in the study and to assure that all modelling-based analysis is reproducible and defensible. Good practice in model development and application deserves close attention. In fact, the credibility and impact of the information and insight that modelling aims to generate are highly dependent on the quality of the modelling exercise. "*It is crucial for model acceptance and is a necessity for long term, systematic accrual of a good knowledge base for both science and decision-making. The complexity and uncertainty inherent in management for better sustainability outcomes make the pursuit of good practice especially important, in spite of limited time and resources*" Jakeman et al. (2006: p. 602).

The aspect of good practice in modelling can become a real problem when modelling moves from the research to the management domain. Managers and interest groups are not scientists working in sharply defined areas of research, but people generally pressurized to use models to manage complex situations. They usually do not have a modelling or quantitative background, and therefore find it difficult to judge their quality or appropriateness, as the previous authors observed.

The conflict of interest among actors is another critical issue that is well identified by McGlade (2002: p. 4) "*Significantly, these conflicts operate at local, regional, national and European scales and reflect fundamental differences in perception and value systems. Thus a crucial issue, for any conception of sustainable management, is the need to understand the socio-environmental driving forces of change at different spatio-temporal scales. What this means is an ability to assess the resilience of socio-natural landscapes to a variety of human and naturally induced pressures - effectively, developing an understanding of the variable sensitivities of ecological, economic and socio-cultural processes, so as to anticipate likely future outcomes and possible unforeseen evolutionary trajectories*".

That water management is closely related to political processes is well clarified by Pahl-Wostl (2006: slide 2), who suggests that *"Adaptive management is needed as a systematic process for continually improving management policies and practices by learning from the outcomes of implemented management strategies."*

There is great need for a new approach, which could progress from traditional, purely quantitative modelling towards a multilevel approach, as Sendzimir et al. (2007) recommend. At a higher level, the definition of a common conceptual framework for cross-disciplinary work involving authorities, managers, stakeholders and researchers and this task can be achieved by conceptual models. Such systemic models identify components (variables) and clarify structures (web of interactions, feedback loops and delays), but, operating in qualitative terms, they do not quantify the implications of different policies or processes on system dynamics. At lower levels, quantitative models can support specific analysis, exploring the strength of interactions and the sensitivity of the system to change. The information produced, clarifying the areas of greatest uncertainty and influence on the system's evolution under different assumptions, can help prioritize field research and support water and agricultural management.

This approach may be particularly useful for implementing the WFD in the agricultural sector, where analyses at different scales are needed to quantify aggregate impacts at the basin scale, such as non-point pollution or water demand, and to design and assess alternative packages of measures, both in agricultural and water, addressing specific problems at farm and field scales.

Even if in recent years important progress has been made in modelling and tools, this is not enough to guarantee that tools are able to effectively support implementation of the WFD at the operational level across Europe, at least in the agricultural sector.

The recent reform of the Common Agricultural Policy is perhaps a significant barrier, since many measures, such as the decoupling of subsidies, eco-conditionality, agro-environmental schemes, are new. Furthermore, the measures to implement such policies are very locality-dependent and, in many cases, have not yet been defined. What emerges from this review is that many existing tools are unable to address these policies. There is a clear need for new tools to conduct an ex-ante assessment of agricultural and environmental policies and technological innovations. The integrated framework SEAMLESS-IF, with its multi-scale capabilities ranging from field and farm scales to the EU25 and the world, including models that simulate effects on the environment and on economic developments, besides procedures that enable the assessment of the social aspects of sustainability, could represent an important tool in the near future to address agricultural policies.

The economic analysis requirement of the WFD, which addresses different aspects ranging from the economic analysis of water use, cost recovery and pricing schemes, to the evaluation of benefits to the environment and society and eventually of disproportionate costs, introduces further complexity for modelling and tools. This is because the methodological framework has not yet been defined. In fact, a multiplicity of approaches, methods and techniques exist to address specific aspects of the monetary or non-monetary valuation, but no common guidelines are available. Even if many tools exist, no one seems capable of fully addressing the economic and policy aspects related to the implementation of the WFD in agriculture. Distinct aspects seem relevant, as far as economic analysis is concerned:

- The supply cost can be assessed by traditional financial approaches implemented in available accounting programs.
- The valuation of environmental cost and benefit (externalities) should be addressed by proper economic techniques, requiring a high level of scientific expertise.
- Resource costs could instead benefit modelling approaches that are capable of considering alternative water allocation under scarcity. Such models are not easily available to final users, even if there have been positive experiences in experimental case studies, which are usually elements of research projects. In any case, use of such models requires a high level of specific competencies.

A different type of model is required when the goal is to ex-ante assess the multidimensional impact of alternative measures at regional/sub-catchment levels.

The integration of existing models in an 'Open Modelling Interface and Environment', such as in the projects described in this document, could represent an innovative support for users.

At present, the integration of different tools, as the case studies show, can improve transparency in decision-making and favour the comparison of different management options in terms of their ecological, economic and social impact. All of the described tools may be beneficial to this process, if used properly. However, the choice of tool is highly dependent on local circumstances, among which previous experience in modelling, trained staff and data availability are critical.

The translation of the experience gained in recent research to the management domain will probably be a long process, in which support by the scientific community should be viewed as essential. The following recommendations are considered important to bridge the gap to real-world decision processes:

- a multilevel approach in modelling and integrating conceptual and quantitative models should be adopted;
- participatory modelling involving end users and stakeholders in the modelling exercises should be implemented in order to enable them to understand and review the various assumptions and their implications for the modelling results;
- training and capacity-building activities are necessary at the local level;
- models and tools must be adapted to pre-established local approaches, and not vice versa;
- models and tools should be seen as ICT components of a structured, flexible approach.

The suggested approach could solve the problem at the national level, favouring an enforceable, transparent, reproducible, auditable process. However, this could conflict with the European dimension. Nationally based approaches, which vary throughout Europe, could create a situation in which the resulting assessments and decisions are not fully comparable between countries.

Since innovation in computational technology will further enhance the frontier of modelling, dedicated web portals offering comprehensive, shared European information, such as WISE 'Water Information System for Europe', are a key instrument for water and agricultural managers and other interested parties to access updated information.

8. REFERENCES

Acutis, M., Ducco, G., Grignani, C., 2000. Stochastic use of LEACHN model to forecast nitrate leaching in different maize cropping systems, European Journal of Agronomy 13, 191–206.

Amador, F., Sumpsi, M., Romero, C., 1998. A non interactive methodology to assess farmers' utility functions: an application to large farms in Andalusia, Spain, European Review of Agricultural Economics 25, 95-109.

Argent, R.M., 2004. An overview of model integration for environmental applications and components, frameworks and semantics, Environmental Modelling & Software 19, 219-234.

Arnold, J.G., Allen, P.M., Bernhardt, G., 1993. A comprehensive surface groundwater flow model, Journal of Hydrology 142, 47-69.

Bazzani, G.M., 2005a. A decision support for an integrated multi-scale analysis of irrigation: DSIRR, Journal of Environmental Management 77(4), 301-314.

Bazzani, G.M., 2005b. An integrated decision support system for irrigation and water policy design: DSIRR, Environmental Modelling & Software, 20(2), 153-163.

Bazzani, G.M., Di Pasquale, S., Gallerani, V. and Viaggi, D., 2005. Water Framework Directive: Exploring policy design issues for irrigated systems in Italy, Water Policy 7(4), 413-428.

Berbel, J. and Gutierrez, C. (editors), 2004. Sustainability of European Irrigated Agriculture under Water Framework Directive and Agenda 2000, EUR 21220, Luxembourg, Office for Official Publications of the European Communities, 49-67. (ISBN 92-894-8005-X).

Berbel, J., Rodriguez, A. 1998. A MCDM approach to production analysis: an application to irrigated farms in Southern Spain, European Journal of Operational Research 107, 108-118.

Borowski, I., Heeb, J., 2005. Report on 2nd Policy Workshop of Harmoni-CA/WP5: Interaction of Common Agricultural Policy & the implementation of the Water Framework Directive on European and regional level 4/5th April 2005, Brussels. Harmoni-CA Document: HCA-WP5-2005-Re05/Final Version-Deliverable No D 5.3.2 20.04.05.

Cai, X., Rosegrant, M.W., 2004. Optional water development strategies for the Yellow River Basin: Balancing agricultural and ecological water demands, Water Resources Research 40, W08S04, doi:10.1029/2003WR002488.

Di Luzio, M., Srinivasan, R., Arnold, J.G., 2002. Integration of watershed tools and SWAT model into BASINS, Journal of the American Water Resource Association 38 (4), 1127-1142.

Donatelli, M., Stöckle, C.O., Ceotto, E., Rinaldi, M., 1997. Evaluation of CropSyst for cropping systems at two locations of northern and southern Italy, European Journal of Agronomy 6, 35-45.

Doorenbos, J., Kassam, A.H., 1979. Yield response to water, FAO Irrigation and Drainage Paper No. 33. FAO, Rome.

Doppler, W., Salman, A.Z., Al-Karablieh, E.K., Wolff, H., 2002. The impact of water price strategies on the allocation of irrigation water: the case of the Jordan Valley, Agricultural Water Management 55, 171-182.

European Commission, 2003. Explanatory memorandum: a long term perspective for sustainable agriculture. 5586/03 COM (2003) 23 final.

European Commission, 2003. Fact Sheet: Rural Development in the European Union, November 2003. [http://europa.eu.int/comm/agriculture/rur/publi/index_en.htm].

European Commission, 2004. Fact Sheet: New perspectives for EU Rural Development, September 2004. [http://europa.eu.int/comm/agriculture/rur/publi/index_en.htm].

European Commission, 2005a. Commission adopts Strategic Guidelines for Rural Development 2007-2013, Press release, IP/05/845, 05/07/2005, [http://europa.eu.int/rapid/pressReleasesAction.do?reference=IP/05/845&format].

European Commission, 2005b. Proposal for a Council Decision on Community strategic guidelines for Rural Development (Programming period 2007–2013), COM(2005) 304 final 2005/0129, Brussels, 5.7.2005.

European Commission, DG Agriculture, 2004. Impact assessment of rural development programmes in view of post 2006 rural development policy, Final Report, November 2004.

Fedra, K., Jamieson, D.G., 1996. The WaterWare decision-support system for river-basin planning 2. Planning capability, Journal of Hydrology 177, 177-198.

Flichman G., Varela Ortega C., Garrido A., 1995. Agricultural policy and technical choice: a regional analysis of income variation, soil use and environmental effects under uncertainty and market imperfections in environmental and land use issues: An economic perspective. Ed. Luis Miguel Albisu & Carlos Romero, Wissenschaft-verlag Vauk Kiel KG.

Gómez-Limón, J.A., Arriaza, M., Berbel, J., 2002. Conflicting implementation of agricultural and water policy in irrigated areas in the European Union, Journal of Agricultural Economics 53, 259-281.

Hattermann, F., Krysanova, V., Wechsung, F., Wattenbach, M., 2005. Hydrological validation of the ecohydrological model SWIM in a macroscale river basin: from the meso- to the macroscale approach, Hydrological Processes 19, 693-714.

He, C., 2003. Integration of geographic information systems and simulation model for watershed management, Environmental Modelling & Software 18, 809-813.

Heckelei, T., Britz, W., 2005. Models Based on Positive Mathematical Programming: State of the Art and Further Extensions, Plenary paper presented at the 89th EAAE Seminar – 3-5 February 2005 – Parma. Paper available at www.unipr.it/arpa/dipseq/EAAE/PR/Homepage.htm.

Heinz, I., Pulido-Velasquez, M., Lund, J.R., Andreu, J., 2007. Hydro-economic Modeling in River Basin Management: Implications and Applications for the European Water Framework Directive, Springer Netherlands, ISSN 0920-4741 (Print) 1573-1650 (Online).

Howitt, R.E., 1995. Positive mathematical programming, American Journal of Agricultural Economics, 77(2): 329-342.

Iglesias, E., Garrido, A., Gómez-Ramos, A., 2003. Evaluation of drought management in irrigated areas, Agricultural Economics 29, 211-229.

Jakeman, A.J., Letcher, R.A., Norton, J.P., 2006. Ten iterative steps in development and evaluation of environmental models, Environmental Modelling & Software 21, 602-614.

Jakeman, A.J., Letcher, R.A., 2003. Integrated assessment and modelling: features, principles and examples for catchment management, Environmental Modelling & Software 18, 491-501.

Johnsson, H., Larsson, M., Mårtensson, K., Hoffmann, M., 2002. SOILNDB: a decision support tool for assessing nitrogen leaching losses from arable land, Environmental Modelling & Software 17, 505-517.

Krysanova, V., Hattermann, F., Wechsung, F., 2007. Implications of complexity and uncertainty for integrated modelling and impact assessment in river basins, Environmental Modelling & Software 22, 701-709.

Krysanova, V., Wechsung, F., Hattermann, F., 2005. Development of the ecohydrological model SWIM for regional impact studies and vulnerability assessment, Hydrological Processes 19, 763-783.

Lambin, E.F., Rounsevell, M.D.A., Geist, H.J., 2000. Are agricultural land-use models able to predict changes in land-use intensity? Agriculture, Ecosystems and Environment, 82, 321-331.

Letcher R.A., Croke, B.F.W., Jakeman, A.J., 2007. Integrated assessment modelling for water resource allocation and management: A generalised conceptual framework, Environmental Modelling & Software, 22, 733-742.

Lewis, D.R., McGechan, M.B., McTaggart, I.P., 2003. Simulating field-scale nitrogen management scenarios involving fertiliser and slurry applications, Agricultural Systems 76, 159-180.
Matthies, M., Berlekamp, J., Lautenbach, S., Graf, N., Reimer, S., 2006. System analysis of water quality management for the Elbe river basin, Environmental Modelling & Software 21, 1309-1318.
McGlade, J., 2002. Landscape Sensitivity, Resilience and Sustainable Watershed Management: A Co-evolutionary Perspective. Discussion paper prepared for The AQUADAPT workshop, Montpellier October 25th-27th 2002.
McKinney, D.C., Cai, X., Rosegrant, M.W., Ringler, C., Scott, C.A., 1999. Modeling Water Resources Management at the Basin Level: Review and Future Directions. International Water Management Institute, Colombo, Sri Lanka.
Merritt, W.S., Croke, B.F., Jakeman, A.J., Letcher, R.A., Perez, P., 2004. A biophysical toolbox for assessment and management of land and water resources in rural catchments in Northern Thailand, Ecological Modelling 171, 279-300.
Mysiak, J., Camera, R., Sgobbi, A., Giupponi, C., 2006. Facilitating participatory decision processes in integrated river basin management, Communications at the 3rd Harmoni-CA conference, Osnabrueck, 5-7 April 2006, Germany.
Neelamkavil, F., 1988. Computer Simulation and Modelling, John Wiley and Sons, Chichester.
Pahl-Wostl, C., 2002. Participative and stakeholder-based policy design and modeling processes, Integrated Assessment 3, 3-14.
Pahl-Wostl, C., 2006. Adaptive management and social learning: The need for a transition in water management and governance, World Water Forum, Mexico, March 2006 Harmoni-CA Forum & Conference Osnabrueck, 5-7 April 2006, Germany.
Pearce, D.W., Turner, R.K., 1990. Economics of Natural Resources and the Environment, Harvester Wheatsheaf, Hemel Hempstead.
Peralta, J.M., Stöckle, C.O., 2001. Dynamics of nitrate leaching under irrigated potato rotation in Washington State: a long-term simulation study, Agriculture, Ecosystems and Environment 88, 23-34.
PIK, 2003. ArcEGMO Documentation. Potsdam Institute of Climate Impact Research (PIK) http://www.arcegmo.de/public/index.html
Quinn, N.W.T., Brekke, L.D., Miller, N.L., Heinzer, T., Hidalgo, H., Dracup, J.A., 2004. Model integration for assessing future hydroclimate impacts on water resources, agricultural production and environmental quality in the San Joaquin Basin, California, Environmental Modelling & Software 19, 305-316.
Reca, J., Roldan, J., Alcaide, M., Lopez, R. and Camacho, E., 2001. Optimisation model for water allocation in deficit irrigation systems. I. Description of the model, Agricultural Water Management 48, 103-116.
Refsgaard, J.C. (Ed.), 2005. State-of-the-Art Report on Quality Assurance in Modelling Related to River Basin Management http://www.harmoniqua.org.
Roetter, R.P., van den Berg, M., Laborte, A.G., Hengsdijk, H., Wolf, J., van Ittersum, M., van Keulen, H., Agustin, E.O., Son, T.T., Lai, N.X., Guanghuo, W., 2007. Combining farm and regional level modelling for integrated resource management in East and South-east Asia, Environmental Modelling & Software 22, 149-157.
Romero, C., Rehman, T., 1989. Multiple Criteria Analysis for Agricultural Decision, Amsterdam, Elsevier.

Röpke, B., Bach, M., Frede, H.G., 2004. DRIPS-a decision support system estimating the quantity of diffuse pesticide pollution in German river basins, Water Science and Technology 49(3), 149-56.

Rosegrant, M.W., Ringler, C., McKinney, D.C., Cai, X, Keller, A., Donoso, G., 2000. Integrated economic-hydrologic water modeling at the basin scale: The Maipo river basin, Agricultural Economics 24, 33-46.

Roy, B., 1996. Multicriteria Methodology for Decision Aiding. Kluwer Academic Publishers, Dordrecht.

Saaty,T.L.,1980.The Analytic Hierarchy Process. McGraw Hill, New York.

Sendzimir, J., Magnuszewski, P. Balogh, A., 2007. Anticipatory modeling of biocomplexity in the Tisza River Basin: First steps to establish a participatory adaptive framework, Environmental Modelling & Software 22, 599-609.

Stöckle, C.O., Debaeke, P., 1997. Modeling crop nitrogen requirements: A critical analysis, European Journal of Agronomy 7, 161–169.

Sumpsi, J.M., Amador, F., Romero, C., 1996. On farmers' objectives: A multicriteria approach, European Journal of Operational Research 96, 64-71.

Ten Berge, H.F.M., van Ittersum, M.K., Rossing, W.A.H., van de Ven, G.W.J., Schans, J., van de Sanden, P.A.C.M., 2000. Farming options for The Netherlands explored by multi-objective modelling, European Journal of Agronomy 13, 263–277.

Tiwari, D.N., Loof, R., Paudyal, G.N., 1999. Environmental decision-making in lowland irrigated agriculture using multi-criteria analysis techniques, Agricultural Systems 60, 99-112.

Varela-Ortega, C., Sumpsi, J.M., Garrido, A., Blanco, M., Iglesias, E., 1998. Water pricing policies, public decision making and farmers response: Implications for water policy, Agricultural Economics 19, 193-202.

Ward, F., Lynch, T., 1996. Integrated river basin optimization: Modelling economic and hydrologic interdependence, Water Resources Bulletin 32, 1127–1138.

Ward, F.A., Michelsen, A., 2002. The economic value of water in agriculture: Concepts and policy applications, Water Policy 4, 423-446.

Williams, J.R., Renard, K.G., Dyke, P.T., 1984. EPIC: A new model for assessing erosion's effect on soil productivity, Journal of Soil and Water Conservation 38, 381-383.

Wolf, J., Beusen, A.H.W., Groenendijk, P., Kroon, T., Rötter, R., van Zeijts, H., 2003. The integrated modeling system STONE for calculating nutrient emissions from agriculture in the Netherlands, Environmental Modelling & Software 18, 597–617.

Yang H, Zhang X., Zehnder, A., 2003. Water scarcity, pricing mechanism and institutional reform in northern China irrigated agriculture, Agricultural Water Management 62, 1-19.

Yaron, D., Dinar, A., 1982. Optimal allocation of farm irrigation water during peak seasons, American Journal of Agricultural Economics 64, 681-689.

9. ANNEXES

9.1 Annex 1 EU-funded projects related to water and agricultural modeling

Table 7 EU-funded projects related to water and agricultural modelling

Title	Acronym
FP VI	
A Regional Model for Integrated Water Management in Twinned River Basins	RIVERTWIN
Development and Testing of Practical Guidelines for the Assessment of Environmental and Resource Costs and Benefits in the WFD	AQUAMONEY
Groundwater Artificial recharge Based on Alternative sources of wateR: aDvanced INtegrated technologies and management	GABARDINE
Integrated decision support system for risk assessment and management of the water-sediment-soil system at river basin scale in fluvial ecosystems	RAMWASS
Integrated Project to Evaluate the Impacts of Global Change on European Freshwater Ecosystems	EURO-LIMPACS
Mitigation of Water Stress through new Approaches to Integrating Management, Technical, Economic and Institutional Instruments	AQUASTRESS
New Approaches to Adaptive Water Management under Uncertainty	NEWATER
Participatory multi-Level EO-assisted tools for Irrigation water management and Agricultural Decision-Support	PLEIADeS
Science-Policy Interfacing in support of the Water Framework Directive implementation	Spi-Water
Understanding river-sediment-soil-groundwater interactions for support of management of waterbodies (river basin & catchment areas)	AQUATERRA
Water scenarios for Europe and for neighbouring states	SCENES
System for Environmental and Agricultural Modelling; Linking European Science and Society	SEAMLESS

(continued)

Table 7 *Continued*

Title	Acronym
FP V	
A decision support system to quantify cost/benefit relationships of the use of vegetation in the management of heavy metal polluted soils and dredged sediments	PhytoDec
Afforestation management in North-Western Europe - influence on nitrogen leaching, groundwater recharge, and carbon sequestration	AFFOREST
An environmental soil test to determine the potential from sediment and phosphorus transfer in run-off from agricultural land	DESPRAL
Analysis, synthesis and transfer of knowledge and tools on hydrological drought assessment through a European Network	ASTHyDA
ARID CLUSTER: Strengthening complementarity and exploitation of results of related RTD projects dealing with water resources use and management in arid and semi-arid regions	ARID
Benchmark models for the Water Framework Directive	BMW
Computer aided rehabilitation of water networks	CARE-W
Data assimilation within a unifying modeling framework for improved river basin water resources management	DAUFIN
Developing strategies for regulating and managing water resources and demand in water deficient regions	WaterStrategyMan
European water management between regulation and competition	AQUALIBRIUM
European water regimes and the notion of a sustainable status	EUWARENESS
Evaluation and improvement of water quality models for application to temporary waters in southern European catchments	TempQsim
Evaluation of alternative techniques for determination of water budget components in water-limited, heterogeneous land-use systems	WATERUSE
Floodplain biodiversity and restoration 2: integrated natural science and socio-economic approaches to catchment flow management	FLOBAR2
Freshwater integrated resource management with agents	FIRMA
Guidelines for the organisation, use and validation of information systems for evaluating aquifer resources and needs	GOUVERNE

(*continued*)

Table 7 *Continued*

Title	Acronym
FP V	
Harmonised modelling tools for integrated basin management	HarmoniCA
Harmonised techniques and representative river basin data for assessment and use of uncertainty information in integrated water management	HarmoniRiB
Harmonising collaborative planning	HarmoniCOP
Harmonising quality assurance in model-based catchment and river basin management	HarmoniQuA
Human effects on nutrient cycling in fluvial ecosystems: development of an expert system to assess stream water quality management at reach scale	STREAMES
Integrated assessment tools to gauge local functional status within freshwater ecosystems	TARGET
Integrated evaluation for sustainable river basin governance	ADVISOR
Integrated nitrogen model for European catchments	INCA
Integrated soil and water protection	SOWA
Integrated strategies for the management of transboundary waters on the Eastern European fringe - the pilot study of lake Peipsi and its drainage basin	MANTRA-East
Integrated water management of transboundary catchments	TRANSCAT
Integrated water resources management for important deep European lakes and their catchment areas	EUROLAKES
IT frameworks	HarmonIT
Multi-sectoral, integrated and operational decision support system for sustainable use of water resources at the catchment scale	MULINO
Management of the environment and resources using integrated techniques	MERIT
MONITOR Water management system based on innovative monitoring equipment and DSS	WATER
Nutrient management in the Danube basin and its impact on the Black Sea	DANUBS
Pesticides in European groundwaters: detailed study of representative aquifers and simulation of possible evolution scenarios	PEGASE

(*continued*)

Table 7 *Continued*

Title	Acronym
FP V	
RATE An advanced study course in water resources management in arid semi-arid regions using advanced earth observation and GIS techniques	WATER
Real-time flood decision support system integrating hydrological, meteorological and radar technologies	FLOODRELIEF
RIVER PROJECT The Tisza river project - real-life scale integrated catchment models for supporting water- and environmental management decisions	TISZA
Social learning for the integrated management and sustainable use of water at catchment scale	SLIM
Stochastic analysis of well head protection and risk assessment	W-SAHARA
Strategic tools to support adaptive, integrated water resources management under changing conditions at catchment level: a co-evolutionary approach	AQUADAPT
Sustainable water: uncertainty, risk and vulnerability in Europe	SWURVE
Systems approach to environmentally acceptable farming	AGRIBMPWATER
The sustainability of European irrigated agriculture under Water Directive and Agenda 2000	WADI
Towards harmonised procedures for quantification of catchment scale nutrient losses from European catchments	EUROHARP
Towards sustainable water use on Mediterranean islands: addressing conflicting demands and varying hydrological, social and economic conditions	MEDIS
Water observing and information system for decision support	WOISYDES
Web-based European knowledge network on water	WEKNOW

9.2 Annex 2: The Common Agricultural Policy

The Common Agricultural Policy (CAP) has a long history that goes back to the Treaty of Rome, the founding document of what has become the European Union (EU), signed in 1957[33].

[33]The Treaty of Rome was signed by France, West Germany, Italy, the Netherlands, Belgium and Luxembourg. The treaty identified the following objectives, among others: to increase agricultural productivity; to ensure a secure food supply at reasonable prices; and to give the agricultural community a fair income. These aims were to be achieved through a free internal market with high domestic prices. The CAP is one of the main European policies, at least in terms of resources.

A variety of driving forces, which underlie farmers' management decisions, influenced this process: technological development, changes in market conditions, changes in the costs of labour, land and other factors of production, modification of the socio-cultural environment, structural changes and a range of different public policies, including the CAP itself[34].

There is evidence that the effect of this intervention has been multi-fold and sometimes contradictory: a clear conflict between agricultural and environmental goals has emerged. Production subsidies in the form of price support to agricultural commodities have generally encouraged more intensive production patterns based largely on machinery, genetic seed, industrial fertilizers and pesticides, which increased yields, but also environmental pressures. Water demands (direct effect) and surface water and groundwater pollution from the run-off of agricultural chemicals (indirect effect) are both relevant in this context.

Since its establishment, the whole concept of the CAP and the way in which it works have greatly changed, due to successive reforms. Environmental integration into other EU policies, including the CAP, is part of a long process dating back to the 1980s. It is legally bound in the Treaty of the European Community (TEU): "Environmental protection requirements must be integrated into the definition and the implementation of (all) the Community policies and activities . . . in particular with a view to promoting sustainable development[35]". The Agreement on the Mid-Term Review of the CAP, reached in June 2003, gives a clear shape to a new European model for agriculture in the years ahead, reflecting the multifunctional[36] role that farming plays, and integrating economic viability, food safety, social balance and environmental concerns.

Reg. 1782/2003[37], the reference document for the reform, defines:

---

[34]National and regional economic, social, environmental, fiscal and land use policies can influence farmers' decisions.

[35]The 5th Environmental Action Programme refers to sustainable development as "development which meets the needs of the present without compromising the ability of future generations to meet their own needs". This entails preserving the overall balance and value of the natural capital stock and taking a long-term view of the real socio-economic costs and benefits of consumption and conservation. The achievement of "balanced and sustainable development", together with the promotion of economic and social progress and a high level of employment, is the first overall objective of the Union (Art. 2 of the TEU).

[36]The term conveys that agriculture provides a variety of non-marketable outputs and services of good public character, which are valuable to the well-being of society.

[37]Council Regulation (EC) No. 1782/2003 of 29 September 2003 establishing common rules for direct support schemes under the Common Agricultural Policy and establishing certain support schemes for farmers and amending Regulations (EEC) No. 2019/93, (EC) No. 1452/2001, (EC) No. 1453/2001, (EC) No. 1454/2001, (EC) No. 1868/94, (EC) No. 1251/1999, (EC) No. 1254/1999, (EC) No. 1673/2000, (EEC) No. 2358/71 and (EC) No. 2529/2001.

- "Single payment scheme": income support decoupled from production (Art. 1) based on entitlements over the 2000-02 reference period (Art. 33) is maintained to provide farmers with a fair income. Farmers shall receive a payment entitlement per hectare, which is calculated by dividing the reference amount by the three-year average number of all hectares entitling them to direct payments in the reference period (Art. 43). Any payment entitlements accompanied by eligible hectares[38] shall entitle the farmer to the payment (Art. 44)[39]. Member States shall constitute a national reserve of payment entitlements to assist new farmers. Premium rights can only be transferred under certain conditions; transfers will support the national reserve (Art. 42), which could be used in order to avoid land being abandoned and the maintenance of rural areas[40].
- "Cross-compliance"[41]: direct payments to farmers are made subject to certain requirements (Art. 3) in the following areas: public, animal and plant health, the environment and animal welfare (Art. 4). Requirements, however, are not defined on a Community-wide basis but in accordance with criteria set by the Member States[42]. In particular, Member States are obliged to define at national or regional levels, minimum requirements for good agricultural and environmental conditions, taking into account the specific characteristics of the areas concerned, including soil and climatic conditions, existing farming systems, land use, crop rotation, farming practices and farm structures (Art. 5). Member States have to set up a "farm advisory system[43]" by 1 January 2007, advising farmers on land and farm management (Art. 13), and implement control systems to ensure compliance with the statutory management

[38]"Eligible hectare" shall mean any agricultural area of the holding taken up by arable land and permanent pasture except areas under permanent crops, forests or used for non-agricultural activities (Art. 44). In order to avoid distortions of competition- fruit and vegetables, including potatoes, are excluded from production on eligible land expect, when regionalisation is implemented.

[39]"Set-aside entitlements" are also implemented (Art. 53) following Art. 6(1) of Regulation (EC) No. 1251/1999.

[40]The CAP also adopts a "financial discipline", with a view to ensuring that the amounts for financing the CAP respect annual ceilings (Art. 11). A maximum guaranteed area should be prescribed and proportional reductions applied if this area is exceeded, concentrated in Member States that overshoot their area.

[41]Commission Regulation (EC) No. 796/2004 of 21 April 2004, which lays down detailed rules for the implementation of cross-compliance, modulation and the integrated administration and control system provided for in Council Regulation (EC) No. 1782/2003, establishing common rules for direct support schemes under the Common Agricultural Policy and establishing certain support schemes for farmers.

[42]For new Member States, access to the new single payments is not necessarily subject to conditionality. Commission Regulation (EC) No. 795/2004 of 21 April 2004 lays down detailed rules for the implementation of the single payment scheme [Official Journal L 141 of 30.04.2004].

[43]The system should help commercial farmers to become more aware of material flows and on-farm processes relating to the environment, food safety, animal health and welfare.

requirements and good agricultural and environmental conditions (Art. 25). Where it is found that farmers do not comply with the eligibility conditions, the payment or part thereof shall be subject to reductions and exclusions, graduated according to the severity, extent, permanence and repetition of non-compliance (Art. 24).

- Rural development, which forms the "second pillar" of the CAP and aims to set new political objectives, is strengthened[44]. All instruments linked to RD policy are grouped into one single instrument, designed to:
  - increase the competitiveness of the agricultural sector through support for restructuring;
  - enhance the environment and countryside through support for land management, including co-financing rural development actions related to Natura 2000 nature protection sites; and
  - enhance the quality of life in rural areas, promoting the diversification of economic activities through measures targeting the farm sector and other rural actors.

In order to achieve a better balance between policy tools designed to promote sustainable agriculture by market policies, the "first pillar" and those designed to promote RD (a system of progressive reduction of direct payments), is introduced on a compulsory Community-wide basis "Modulation" (Art. 10)[45].

9.3 Annex 3: Measures to implement the WFD and the CAP

The programmes of measures can be considered as the principal mechanism for the implementation of the environmental objectives of the WFD, and have to be established by 2009 and made operational by 2012 (Art. 11 WFD). They should be based on risk assessment (Art. 5 WFD). The WFD distinguishes between basic measures (minimum requirements), including the so-called combined approach for point and diffuse sources (Art. 10 WFD), and supplementary measures. The former include the implementation of a number of environmental directives that directly or indirectly assist in the protection of water (Annex VI, Part A WFD). On the agricultural side, 'cross compliance' links the CAP to European environmental directives (Annex III) by 1.1.2005. Some of these appear in both lists: the conservation of wild birds; the protection of the environment, and in

[44]Council Regulation (EC) No. 1698/2005 of 20 September 2005 on support for rural development by the European Agricultural Fund for Rural Development (EAFRD).

[45]All direct payment amounts to be granted in excess of EUR 5000 shall be reduced for each year until 2012 by the following percentages: 2005: 3%, 2006: 4%, 2007-12: 5%. The annual budget allocated to the first pillar over this period is EUR 40.5 billion. It has also allotted EUR 14 billion to RD and to veterinary and plant health measures.

particular of the soil, when sewage sludge is used in agriculture; the protection of waters against pollution caused by nitrates from agricultural; the conservation of natural habitats and of wild flora and fauna[46]. This creates common ground for both policies[47].

River basin programmes of measures can also include ad hoc measures. In this regard, the design of upcoming RD programmes (2007-2013) is an opportunity to support the WFD in the agricultural sector. Axis 2 concerning "Environment and land management" can be used to devise directly targeted measures. However, other measures concerning "Improving competitiveness of farming and forestry" (Axis 1) and "Improving quality of life and diversification" (Axis 3) offer opportunities to contribute indirectly to WFD delivery.

The Common Agricultural Policy's agro-environmental programmes are often subdivided into different schemes, each made up of a series of measures. Agricultural and RD policies include a rich set of tools. A basic distinction is made between "Good Farming Practices" (GFP) and "Agri-Environment Schemes" (AES), aiming to reach different environmental quality levels. Member States have to specify verifiable standards representing baselines to be respected for both measures. The reference level can be used to distinguish between positive and negative environmental effects. GFP should ensure that farmers observe mandatory environmental requirements in line with the Polluter Pays Principle, and therefore aim to prevent negative environmental effects (e.g. water pollution). Penalties should normally apply in enforcing the avoidance of negative environmental effects below the reference level of GFP. Conversely, incentives would be offered to reward positive environmental effects above this level. This is the AES case that has to go beyond usual GFP (e.g. a water status higher than the minimum requirement). The scheme provides payments to farmers in return for an environmental service to compensate higher cost and/or lower income which the farmer suffers to comply with the regulation[48].

---

[46]Council Directive 79/409/EEC of 2 April 1979 on the conservation of wild birds (OJ L 103, 25.4.1979, p. 1) Art. 3, 4(1), (2), (4), 5, 7 and 8 2. Council Directive 86/278/EEC of 12 June 1986 on the protection of the environment, and in particular of the soil, when sewage sludge is used in agriculture (OJ L 181, 4.7.1986, p. 6) Art. 3 4. Council Directive 91/676/EEC of 12 December 1991 concerning the protection of waters against pollution caused by nitrates from agricultural sources (OJ L 375, 31.12.1991, p. 1) Art.s 4 and 5. Council Directive 92/43/EEC of 21 May 1992 on the conservation of natural habitats and of wild flora and fauna (OJ L 206, 22.7.1992, p. 7).

[47]Indirect market measures (e.g. food labelling and quality assurance initiatives) can play an important role as a driver to support environmental objectives.

[48]AEM are co-financed by the EU and the Member States with a contribution from the Community budget of 85% in Objective 1 areas and 60% in others. Agri-environment payments are not considered to be trade-distorting subsidies in WTO; they have "Green Box" status.

Annex IV of Reg. 1782/2003 specifies the good agricultural and environmental conditions referred to in Art. 5.

The definition of the reference level may change over time and will vary between Member States and regions according to local environmental conditions, legal traditions and the definition of property rights. AES follow a number of basic principles, which are optional and site-specific, and which require structured and long-term approaches to monitoring and evaluation, and proper payment levels[49]. They are a highly refined tool for environmental integration. They are able to achieve certain environmental results that other instruments are unable to achieve. However, in order to be implemented correctly, they must be enforced, and the existing administrative capacity of Member States must be upgraded.

Table 8 Good agricultural and environmental practice (Annex IV of Reg. 1782/2003, Art. 5 of Council Reg. 1782/2003)

Issue	Standards
Soil erosion: protect soil through appropriate measures	— Minimum soil cover — Minimum land management reflecting site-specific conditions — Retain terraces
Soil organic matter: maintain soil organic matter levels through appropriate practices	— Standards for crop rotations where applicable — Arable stubble management
Soil structure: maintain soil structure through appropriate measures	— Appropriate machinery use
Minimum level of maintenance: ensure a minimum level of maintenance and avoid the deterioration of habitats	— Minimum livestock stocking rates or/ and appropriate regimes — Protection of permanent pasture — Retention of landscape features — Avoidance of the encroachment of unwanted vegetation on agricultural land

[49]Agri-environmental contracts compete economically with the most profitable land use, so payment levels have to be set sufficiently high to attract farmers to join schemes, while avoiding over-compensation. This requires the calculation of appropriate payment levels.

Table 9 Monetary valuation

Methods	Assumption
Market	Prices for goods and services traded in markets capture environmental values.
Cost-based valuation	Costs of maintaining an environmental benefit is a reasonable estimation of preventive and/or mitigation measures[50].
Revealed preference	Environmental values can be indirectly inferred via econometric approaches from observed behaviours. This group includes: – Recreational Demand Methods (RDM): recreation is a function of the quality of water. Improvements to or deterioration in good (water) quality may enhance or reduce recreational opportunities (e.g. number of visits in a given period). – Hedonic Pricing Models (HPM) require the existence of a linked market (e.g. real estate that capture environmental values). – Averting Behaviour Models (ABM) are when people adopt defensive behaviour in response to environmental damage (e.g. new filter for water).
Stated preference	– Values[51] are inferred directly, eliciting consumer preference on either hypothetical or experimental markets. Methods include: – Contingent Valuation (CV). – Contingent Ranking (CR).

9.4 Annex 4: Monetary valuation

Monetary valuation of environmental goods and services has a long tradition where different methods are developed and applied to put monetary values on environmental services via economic models. Table 9 (adapted from Pearce and Turner, 1990) summarizes the characteristics of the available methods.

All of the previous methods can support economic analysis in the implementation of the WFD. Some are particularly suitable for the valuation of external costs. It should be clear that they require a case-by-case approach and specific application in order to capture costs and benefits accurately.

Environmental values can also be inferred through meta-analysis from existing studies. This method is known as value transfer or benefit/cost transfer. However, in this approach, the risk of potential estimation errors is high. Previous studies must be carried out before this approach can be applied.

[50]This is not always true. In some cases, mitigation may not be possible at all, or on the contrary it could be higher than necessary, leading to an over-estimation of environmental costs. Environmental quality levels, such as the Directives' Objectives, should be taken as references.

[51]The most commonly used measures are willingness to pay WTP to avoid environmental costs or to obtain an environmental benefit, and willingness to accept WTA to support an environmental cost or renounce an environmental benefit. The choice of measure is not neutral - income distributions and levels influence these values.

9.5 Annex 5: A categorisation of tools and models for water management and agriculture

Table 10 describes over 50 existing tools related to water management and agriculture. It provides a starting point in the selection of models for particular applications. The collated information reflects a specific moment in time (2007), and is representative of new versions and capabilities within the models.

The table contains the following information:

1. Tool name
2. Brief description
3. Group
4. Spatial scale
5. Irrigation measures
6. Agricultural measures
7. Economic analysis
8. Examples of measures addressed
9. Link to the website
10. Contact
11. Further characterisation
12. Previous application

The 'Group' code identifies the main core of the tool; 'A' = Agronomic, 'H' = Hydrology, 'E' = Economic. Most information is condensed in the description and characterisation. To synthesise very inhomogeneous data, a few characteristics have been reported as dummies (YES/NO) to identify the capability to analyse 'irrigation measures' and 'agricultural measures'. A code that reflects the economic theory embedded in the tool (MI = 'micro', MA = 'macro', OA = 'other approach') is used to synthesise available information on the economic analysis.

If a cell is left "blank", the model does not seem to have that particular function.

The author apologises for any misrepresentations that may be indicated in the table, and also if he was unable to determine that the model had a particular feature[50].

Additional information on the tool (e.g. data needs, time scale, computation time step, arrangement and costs) is available from the website given in the table or by contacting the reference person or institute.

[52]The author hopes that readers will send him additional information, enabling him to update the information.

Table 10 Information sheet on agricultural and water related model

Tool name	Short description	Group	Spatial scale	Irrigation measures	Agricultural measures	Economic analysis	Type of measures	Web page	Contact	Further characterization	Previous application
AGLINK	A multi-country and commodity economic model	E	EU/State/Region	NO	NO	MA	CAP	www.oecd.org/	OECD Secretariat	AGLINK is a dynamic supply-demand model of world agriculture, developed by the OECD Secretariat in close co-operation with Member countries. The overall design of the model focuses particular attention to the potential influence of agricultural policy on agricultural markets in medium term. AGLINK is a partial equilibrium model, primarily of major OECD commodity markets. AGLINK estimates supply, demand and prices. Non-agricultural sectors are not modelled, and are treated as exogenously to the model.	OCDE
AGNPS - Agricultural Non Point Source Pollution Model	AGNPS is a computer model developed to predict non point source pollutant loadings within agricultural watersheds.	A	Farm/Catchment	YES	YES	OA	Crop rotation, Tillage operations, Fertilization schemes, Pesticide applications, Filter strips	www.geog.uni-hannover.de/phygeo/grass/agnps.html	Ronald L. Bingner rbingner@ars.usda.gov	It contains a continuous simulation, surface runoff model designed to assist with determining BMPs, the setting of TMDLs, and for risk & cost/benefit analyses. It is a joint USDA-Agricultural Research Service and -Natural Resources Conservation Service production.	Many Basins

(*continued*)

Table 10 *Continued*

Tool name	Short description	Group	Spatial scale	Irrigation measures	Agricultural measures	Economic analysis	Type of measures	Web page	Contact	Further characterization	Previous application
AgriBMP-Water	A systems approach to environmentally acceptable farming	E	Farm/Catchment	YES	YES	OA	Farm management and agricultural practices. Water management	www.bordeaux.cemagref.fr/public/agribmpwater/index.html	ramon.laplana@bordeaux.cemagref.fr	This tool aims at providing planners with a grid which would allow a comparison between Best Management Practices (BMPs) in terms of environmental efficiency, economic cost and potential acceptability by farmers. It adopts negociative implementation methods.	France
ANIMO - Agricultural Nutrient Model	Dynamic simplified process-based model to simulate N leaching from the soil surface to ground and surface waters.	A	Farm/Catchment	NO	YES	NA	Farm management and agricultural practices	www.sc.dlo.nl/wateruk/index.htm	J.G. Kroes: j.g.kroes@sc.agrc.nl	ANIMO operates as a multi-layer model with the soil surface and groundwater forming the upper and lower system boundaries, respectively. The main processes simulated by the model are: Nitrogen: mineralization of N, immobilization of NH4, NO3 uptake by plants (excess and limited) nitrification of NH4, denitrification of NO3, volatilization of NH4, NO3 leaching Phosphorus. ANIMO can be linked to any hydrological model for the calculation of fluxes and changes in moisture content.	The Netherlands

(*continued*)

Table 10 *Continued*

Tool name	Short description	Group	Spatial scale	Irrigation measures	Agricultural measures	Economic analysis	Type of measures	Web page	Contact	Further characterization	Previous application
APSIM Agricultural Production System Simulator	A tool to simulate biophysical processes in farming systems	A	Farm	NO	YES	OA	Farm management and agricultural practices	www.apsim.info/apsim/Downloads/	apsim-help @csiro.au	APSIM is structured around plant, soil and management modules. These modules include a diverse range of crops, pastures and trees, soil processes including water balance, N and P transformations, soil pH, erosion and a full range of management controls. APSIM resulted from a need for tools that provided accurate predictions of crop production in relation to climate, genotype, soil and management factor while addressing the long-term resource management issues. It relates to the economic and ecological outcomes of management practices in the face of climate risk.	Australia
AQUASTAT	AQUASTAT is FAO's global information system on water and agriculture	H	Catchment	YES	NO	NA	To be defined case by case	www.fao.org/ag/agl/aglw/aquastat/main/index.stm	Land-and-Water@fao.org	The information system consists of: Databases - Database on African dams; Database on institutions; Database on river sediment yields; Countries and regions - Standardized text by country and by region on the state of water resources and agricultural water use; Climate - A tool to provide climate estimates for the land surface of the globe; Water resources - Review of the statistics of renewable water resources by country; Agricultural water use - Review of agricultural water use by country; Irrigation - Global map of irrigated areas, which is a spatial dataset on areas equipped for irrigation.	Many basins

(*continued*)

Table 10 *Continued*

Tool name	Short description	Group	Spatial scale	Irrigation measures	Agricultural measures	Economic analysis	Type of measures	Web page	Contact	Further characterization	Previous application
AQUATOOL	A generalized DSS for water resource planning and operational management	E	Catchment			OA	Land use and water management	www.upv.es/aquatool/	mapuve @hma. upv.es	The SIMGES Fortran-coded mathematical model performs the simulation of the operational management of the system on a monthly basis. It is responsible for the water allocation to water uses and considers the conjunctive use of surface water and groundwater. Operating policies are defined by the following variables: target, minimum and maximum volumes of reservoirs, inter-reservoir relationships and priorities of use, minimum flow in rivers, flow requirements for hydroelectric plants, targeted water demand for each agricultural, industrial and domestic areas and their demand priorities that are used in the water allocation.	Many basins
BASINS 4 - Better Assessment Science Integrating point and Nonpoint Sources	A multi-purpose environmental analysis system that integrates a GIS, national watershed data, environmental assessment and modeling tools	H	Catchment	YES	YES	OA	Land use, water management	www.epa.gov/waterscience/basins/index.html	Modeling and Information Technology Team Standards and Health Protection Division Office of Science and Technology - U.S. Environmental Protection Agency Mailcode - 4305T - 1200 Pennsylvania Ave., NW Washington, DC 20460	BASINS facilitates examination of environmental information, supporting analysis of environmental systems, and providing a framework for examining management alternatives for watersheds. It runs on a non-proprietary, open source, free GIS system ArcView GIS Base Cartographic, Environmental, Environmental Monitoring, Point Source/Loading data - models: Pollutant Load (PLOAD), Soil and Water Assessment Tool (SWAT), Windows Hydrological Simulation Program-Fortran (Win-HSPF) and Enhanced Stream Water Quality Model (QUAL2) GenScn, GENeration and analysis of model simulation SCeNarios.	Many basins

(*continued*)

Table 10 *Continued*

Tool name	Short description	Group	Spatial scale	Irrigation measures	Agricultural measures	Economic analysis	Type of measures	Web page	Contact	Further characterization	Previous application
CAPRI - Common Agricultural Policy Regional Impact Analysis	Economic model to conduct CAP analysis	E	EU/State/Region	NO	NO	MA	CAP, environmental aspects included	www.agp.uni-bonn.de/agpo/rsrch/capri/capri_e.htm	Wolfgang Britz wolfgang.britz@ilr.uni-bonn.de	CAPRI is an agricultural sector model covering both the whole of EU27 and Norway at regional level (250 regions) and global agricultural markets. The following environmental indicators are covered by CAPRI: - Balances for N,P,K - Emissions of Ammonia, Methan and N2O - Global Warming potentials. CAPRI users both work in research institutions and EU Commission services.	Europe
CREAMS - Chemicals, runoff and erosion from agricultural management systems	It is a model for predicting runoff, erosion, and chemical transport from agricultural management systems	A	Catchment	YES	YES	NA	Cropping systems management	www.wiz.uni-kassel.de/model_db/mdb/creams.html	R.Wayne Skaggs skaggs@eos.ncsu.edu	CREAMS is a field scale model for predicting runoff, erosion, and chemical transport from agricultural management systems. It is applicable to field-sized areas. CREAMS can operate on individual storms but can also predict long term averages (2-50 years). The model must be physically based and not require calibration for each specific application, 2) the model must be simple easily understood with as few parameters as possible and still represent the physical system relatively accurately,	USA

(*continued*)

Table 10 *Continued*

Tool name	Short description	Group	Spatial scale	Irrigation measures	Agricultural measures	Economic analysis	Type of measures	Web page	Contact	Further characterization	Previous application
										3) the model must estimate runoff, percolation, erosion and dissolved and adsorbed plant nutrients and pesticides and, 4) the model must distinguish between management practices. Based on these objectives, since the management practices were usually on a field basis, the size of a field to represent the scale of the model was needed.	
CropSyst Suite - Cropping Systems Simulation Model	CropSyst is a user-friendly, conceptually simple but sound multi-year multi-crop daily time step simulation model	A	Farm	YES	YES	NA	Cropping systems management	www.bsyse.wsu.edu/cropsyst/	Claudio O. Stöckle stockle@wsu.edu	CropSyst is a user-friendly, conceptually simple but sound multi-year multi-crop daily time step simulation model. The model has been developed to serve as an analytic tool to study the effect of cropping systems management on productivity and the environment. The model simulates the soil water budget, soil-plant nitrogen budget, crop canopy and root growth, dry matter production, yield, residue production and decomposition, and erosion. Management options include: cultivar selection, crop rotation (including fallow years), irrigation, nitrogen fertilization, tillage operations (over 80 options), and residue management. The model is currently written in Turbo Pascal 6.0 for MS-DOS using an object oriented programming approach.	Different countries

(*continued*)

Table 10 *Continued*

Tool name	Short description	Group	Spatial scale	Irrigation measures	Agricultural measures	Economic analysis	Type of measures	Web page	Contact	Further characterization	Previous application
CROPWAT - Crop and water	A computer program for irrigation planning and management	A	Farm	YES	YES	NA	Cropping systems management	www.fao.org/ag/agl/aglw/tools.stm	The Water Resources, Development and Management Service of FAO, Viale delle Terme di Caracalla, 00100 Rome, Italy Tel: +(39) 06 57055541, Fax: (39-06) 57056275	CROPWAT is a decision support system developed by the Land and Water Development Division of FAO. It is meant as a practical tool to help agro-meteorologists, agronomists and irrigation engineers to carry out standard calculations for evapotranspiration and crop water use studies, and more specifically the design and management of irrigation schemes. It allows the assessment of production under rainfed conditions or deficit irrigation. Standard crop data are included in the program and climatic data can be obtained for 144 countries through the CLIMWAT-database.	Different countries
DANUBIA decision support system	The GLOWA-Danube project has the objective to develop, to apply and to exploit integrative techniques for the modelling of the water cycle in the Upper Danube catchment.	E	Catchment	YES	YES	OA	To be defined case by case	www.glowa-danube.de/	Wolfram Mauser Department of Earth and Environmental Sciences - Chair for Geography and Geographical Remote Sensing University of Munich Luisenstraße 37 D - 80333 Munich phone: ++49 89 2180 6682	Upon completion, DANUBIA will be able to simulate water-related issues of environmental management under ecological, economical and cultural aspects, such as flood risk and protection, agriculture and water quality and quantity, tourism and water as well as water and climate. It will examine the sustainability of the proposed solution scenarios. DANUBIA will contribute to provide optimal solutions for a sustainable environmental management in large, heterogeneous catchments.	Danube basin

(*continued*)

Table 10 *Continued*

Tool name	Short description	Group	Spatial scale	Irrigation measures	Agricultural measures	Economic analysis	Type of measures	Web page	Contact	Further characterization	Previous application
DRIPS - Drainage Spraydrift and Runoff Input of Pesticides in Surface Waters	A decision support system estimating the quantity of diffuse pesticide pollution in German river basins.	H	Catchment	YES	YES	NA	Pesticides management		bjoern.roepke@agrar.uni-giessen.de	The GIS based decision support system DRIPS has been developed to estimate the predicted environmental concentration (PECsw) of pesticides in surface waters resulting from diffuse sources. PECsw can be calculated on a catchment scale by quantifying the expected mean daily inputs of pesticides via surface runoff, tile drainage and spraydrift for various types of river basins characterized by their daily discharges. DRIPS is fitted with a Graphical User Interface (GUI) to provide easy access to the model algorithms. Model parameters like dose rate, DT50, Koc and date of pesticides application, etc. can be modified by the user in order to generate customized scenarios predicting PECsw for a choice of field crops, orchards and vineyards. Results are available as grid cell maps for the territory of Germany with high temporal and spatial resolution.	German river basin
DSIrr - Decision Support for IRRigation	Agricultural and water policy analysis	E	Farm/Catchment	YES	YES	MI	Cropping systems management, water pricing, quota and markets, taxes, subsidies decoupling, agri-env. schemes	www.bo.ibimet.cnr.it/index.php?id=346&L=3	Guido M. Bazzani g.bazzani@ibimet.cnr.it	DSIrr is support tool for scenario analysis using simulation approaches and optimisation bio-economic models. Scenarios evaluate a full range of water development and management options, and take account of multiple and competing uses of water. Scenario can also describe agricultural policy or other exogenous drivers. A multi-scale modelling	Italy

(*continued*)

Table 10 *Continued*

Tool name	Short description	Group	Spatial scale	Irrigation measures	Agricultural measures	Economic analysis	Type of measures	Web page	Contact	Further characterization	Previous application
										approach is adopted and different types of farms can be aggregated to create a model Dnetwork. The individual models are specialised to address specific questions and reflect heterogeneity between farms. Farm preferences can be described using a multi-attribute approach. Changes in irrigation and farm practices are considered. A set of indicators is quantified to assess socio-economic and environmental impacts at farm and regional level. The DS can be linked to other tools.	
DSS for the Elbe River Water Quality Management	River basin management	E	Catchment	YES	YES	OA	Agricultural practices - Land use, water management	www.riks.nl/projects/Elbe-DSS	S. Kofalk, Bundesanstalt für Gewässerkunde (BfG), Mainzer Tor 1, 56068 Koblenz, Phone: +49 (261) 13065330.	Models, spatial and non-spatial data and analysis tools under a user-friendly GIS-based interface. The tool confronts the decision maker with possible measures as well as multiple management objectives. DSS helps the water managers to formulate policy for river basin management and to take appropriate measures to realise policy objectives. As such it could also be a very useful tool for the implementation of EU-WFD. Furthermore the DSS is especially suited to support participative decision making.	Germany
EPIC - Erosion Productivity Impact Calculator	Assess the effect of soil erosion on productivity. Predict the effects of management decisions	A	Farm	YES	YES	MI	Agricultural practices	www.brc.tamus.edu/epic/	Jimmy Williams williams@brc.tamus.edu	EPIC predict the effects of management decisions on soil, water, nutrient, and pesticide movements and their combined impact on soil loss, water quality, and crop yields for areas with homogeneous soils and management.	Many countries

(*continued*)

Table 10 *Continued*

Tool name	Short description	Group	Spatial scale	Irrigation measures	Agricultural measures	Economic analysis	Type of measures	Web page	Contact	Further characterization	Previous application
	on soil loss, water quality, and crop yields									Weather, surface runoff, return flow, percolation, ET, lateral subsurface flow and snow melt. Water erosion; Wind erosion; N & P loss in runoff, nitrogen leaching; Organic N & P transport by sediment; N & P mineralization, immobilization and uptake; Denitrification; Mineral P cycling; N fixation; Pesticide fate and transport; Soil temperature; Crop growth and yield for over 80 crops; Crop rotations; Tillage, Plant environment control (drainage, irrigation, fertilization, furrow diking, liming); Economic accounting; Waste management (feed yards dairies with or without lagoons).	
FARMIS	FARMIS is a comparative-static process-analytical programming model	E	EU/State/Region	NO	YES	MI	CAP and policy measures	www.fal.de/nn_797028/EN/institutes/BW/research/fieldsactivity/ag1__en.html	Werner Kleinhanß werner.kleinhanss@fal.de	Impacts of the Mid-term Review policy reform on German agriculture can be assessed by using simulation approaches and optimization models based on FADN data. The model network includes different models which are kept current and combined with each other as necessary. The individual models are specialized to address specific questions (for example, regional impacts, farm impacts, market impacts). Comparative-static analysis is conducted with the main purpose is to analyze the sectoral effect of policy measures addressing the farm level.	Germany

(*continued*)

Table 10 *Continued*

Tool name	Short description	Group	Spatial scale	Irrigation measures	Agricultural measures	Economic analysis	Type of measures	Web page	Contact	Further characterization	Previous application
GLEAMS - Groundwater Loading Effects of Agricultural Management Systems	The program simulates water quality events on an agricultural field.	A	Field	YES	YES	NA	Farm level management decisions on water quality	http://arsserv0.tamu.edu/models.htm, http://arsserv0.tamu.edu/nrsu/glmsfact.htm	Kevin King king@brc-sun0.tamu.edu	GLEAMS is a continuous simulation, field scale model. It consists of four major components: hydrology, erosion/sediment yield, pesticide transport, and nutrients. GLEAMS was developed to evaluate the impact of management practices on potential pesticide and nutrient leaching within, through, and below the root zone. It also estimates surface runoff and sediment losses from the field. GLEAMS was not developed as an absolute predictor of pollutant loading. It is a tool for comparative analysis of complex pesticide chemistry, soil properties, and climate.	Numerous climatic and soil regions of the world
GTAP - Global Trade Analysis Project	Quantitative analysis of global economic issues within an economy-wide framework	E	EU/Region	NO	NO	MA	CAP trade and policy	https://www.gtap.agecon.purdue.edu/about/project.asp	Hertel, T.W. Center for Global Trade Analysis 403 West State Street West Lafayette IN 47907-2056 United States	GTAP (Global Trade Analysis Project) is a global network of researchers and policy makers conducting quantitative analysis of international policy issues. GTAP's goal is to improve the quality of quantitative analysis of global economic issues within an economy-wide framework. GTAP offers a variety of products, including: data, models, and utilities for multi-region, applied general equilibrium analysis of global economic issues.	Europe

(*continued*)

Table 10 *Continued*

Tool name	Short description	Group	Spatial scale	Irrigation measures	Agricultural measures	Economic analysis	Type of measures	Web page	Contact	Further characterization	Previous application
ICMS - Interactive Component Modelling System	Integrated Scenario Modelling for Managing Catchments	E	Catchment	YES	YES	MI	Agricultural practices - Infrastructure investment, Farm Dams, Environmental Flows, Sleeper licence activation	www.clw.csiro.au/products/icms/	Letcher, R.A. Rebecca @cres.anu.edu.au	The DSS is a framework in which to embed models and tools for analysis and presentation of environmental options for environmental managers. It applies scenarios as model inputs which take into account achievable policy/management drivers and socioeconomic capacity for change, as well as uncontrollable system shocks. The DSS computes, as model outputs, indicators covering a spectrum of biophysical and socioeconomic impacts for each scenario.	Namoi River
IRRINET	The web system entails a complete dissemination of information in agro-meteorology	A	Farm/Catchment	YES	YES	NA	Agricultural practices	www.consorziocer.it	Canale Emiliano Romagnolo cer@consorziocer.it	Through a channel from the Po river water is made available over a 3000 Km2 area and the water resources (68m3/s) allow to satisfy the irrigation needs. Water is furnished together with the information for its rational and sustainable utilization. The web system entails a complete dissemination of information in agrometeorology up to a personalized irrigation guided scheduling for farmers based on water balance models at plot level. The service is active from 1985 and on average favour a reduction of 20 % in water consumption.	Italy

(*continued*)

Table 10 *Continued*

Tool name	Short description	Group	Spatial scale	Irrigation measures	Agricultural measures	Economic analysis	Type of measures	Web page	Contact	Further characterization	Previous application
IRRINET	The web system entails a complete dissemination of information in agro-meteorology	A	Farm/Catchment	YES	YES	NA	Agricultural practices	www.consorzioocer.it	Canale Emiliano Romagnolo cer@consorziocer.it	Through a channel from the Po river water is made available over a 3000 Km2 area and the water resources (68m3/s) allow to satisfy the irrigation needs. Water is furnished together with the information for its rational and sustainable utilization. The web system entails a complete dissemination of information in agrometeorology up to a personalized irrigation guided scheduling for farmers based on water balance models at plot level. The service is active from 1985 and on average favour a reduction of 20 % in water consumption.	Italy
IWRAM - Integrated Water Resource Assessment and Management	An integrated modelling toolbox for considering impacts of development and land use	E	Farm/Catchment	YES	YES	MI	Farm management, development and land use	icam.anu.edu.au/projects/water_management.html	Tony.Jakeman@anu.edu.au	The IWRAM contains two modelling toolboxes utilizing a nodal network structure for catchment analysis: a Biophysical Toolbox (erosion, streamflow, crop) and an Integrated Modelling Toolbox, which links models of household decision making to allow for consideration of socioeconomic and environmental trade-offs of many development and policy scenarios.	Mae Chaem catchment in Northern Thailand.
LADSS - Land Allocation Decision Support System	LADSS is a computer based rural land use planning tool	E	Farm		YES	OA	Land use stategic planning	www.macaulay.ac.uk/LADSS/	http://www.macaulay.ac.uk/	LADSS is a farm-scale land use planning tool being developed within the Integrated Land Use Systems group of The Macaulay Institute in order to assist in the case-based investigation of policy and environmental change impacts on land-use systems.	United Kingdom

(*continued*)

Table 10 *Continued*

Tool name	Short description	Group	Spatial scale	Irrigation measures	Agricultural measures	Economic analysis	Type of measures	Web page	Contact	Further characterization	Previous application
										The tool supports strategic, farm-scale, land use planning by suggesting possible combinations of land uses to meet multiple-objectives· Provides a framework within which the financial, social and environmental consequences of changes in land use may be evaluated.· Demonstrates an approach to the integrating of spatial data and analysis methods with land use systems models· Acts as a channel for technology transfer from land use scientists to land managers and facilitates the inclusion of practitioner knowledge into models of land use systems.	
LEACHM - Leaching Estimation and Chemistry Model	LEACHM refers to a suite of simulation models describing the water and chemical regime in the soil root zone.	H	Farm/Catchment	YES	YES	NA	Agicultural practices	eco.wiz.uni-kassel.de/model_db/mdb/leachm.html	john.hutson@es.flinders.edu.au	The LEACHM suite consists of four simulation models and several utilities. The simulation models utilize similar numerical solution schemes to simulate vertical water and chemical movement. They differ in their description of chemical equilibrium, transformation and degradation pathways. LEACHW describes the water regime only. The other simulations describe pesticides (LEACHP), nitrogen and phosphorus (LEACHN), and salinity in calcareous soils (LEACHC).The models simulate chemical fate and transport in transient-flow	Australia

(*continued*)

Table 10 *Continued*

Tool name	Short description	Group	Spatial scale	Irrigation measures	Agricultural measures	Economic analysis	Type of measures	Web page	Contact	Further characterization	Previous application
										field situations as well as in laboratory columns subject to steady-state or interrupted flow. Water flow and solute transport is described by the Richards and convective-dispersion equations, or by a modified Addiscott mobile/immobile capacity (tipping-bucket) concept.	
Low Flows 2000	A catchment-based water resource decision support tool for the United Kingdom	H	Catchment	YES	YES	NA	Abstraction licensing	www.ceh.ac.uk/products/software/CEHSoftware-LOWFLOWS2000.htm	M.G.R. Holmes: mgrh@ceh.ac.uk	Low Flows 2000 is a decision support tool designed to estimate river flows at ungauged sites and to aid the development of catchment and regional water resources. It is the standard software system used by the Environment Agency and the Scottish Environment Protection Agency for providing estimates of river flows, as represented by annual and monthly flow duration statistics, for any river reach in the UK. The software and underpinning science have been widely published in the scientific literature. The CEH Low Flows 2000 system is now licensed to Wallingford HydroSolutions (WHS) for distribution and further development. WHS was set up in 2004 to maximise the transfer of CEH water research to the user community.	United Kingdom

(*continued*)

Table 10 *Continued*

Tool name	Short description	Group	Spatial scale	Irrigation measures	Agricultural measures	Economic analysis	Type of measures	Web page	Contact	Further characterization	Previous application
MB - Mike Basin	MB is professional engineering software package for integrated river basin planning and management	H	Basin/Catchment	YES	YES	OA	Point and non-point source management alternatives - Land use	www.crwr.utexas.edu/gis/gishyd98/dhi/mikebas/Mbasmain.htm	Danish Hydraulic Institute (DHI) Lilla Bommen 1 S-411 04 Göteborg Sweden Phone: + 46 31 80 87 90	The DSS integrates GIS, data bases and models to provide an user-friendly tool, capable of evaluating alternative options for compliance considering the legislative requirements, technical options for improvements, environmental impacts and economic/financial implications. MIKE BASIN provides a mathematical representation of the river basin encompassing the configuration of the main rivers and their tributaries, the hydrology of the basin in space and time, existing as well as potential major schemes and their various demands of water. The natural river system is schematised and represented with a node-branch structure.	Many basins
MicroLEIS DSS - A Land Evaluation Decision Support System for Agricultural Soil Protection	A computer-based set of tools for an orderly arrangement and practical interpretation of land resources/agricultural management data.	A	Farm/Catchment	YES	YES	NA	Agricultural management, Land resources	leu.irnase.csic.es/microlei/manual2/overview.htm	Prof. D. de la Rosa - IRNAS, Instituto de Recursos Naturales y Agrobiología de Sevilla Avenida de Reina Mercedes, 10. 41012 Sevilla - España. Teléfono: 954624711 Fax: 954624002	MicroLEIS is a agro-ecological system. It major characteristics are: data and knowledge engineering through the use of a variety of databases and innovative modelling techniques; scaling-up of process knowledge from the micro-scale to the landscape-scale; land evaluation; use of monthly meteorological data and standard information; integrated agro-ecological approach, combining bio-physical data with agricultural management experience; software development for PC platforms and Web- and GIS-based versions.	Spain

(*continued*)

Table 10 *Continued*

Tool name	Short description	Group	Spatial scale	Irrigation measures	Agricultural measures	Economic analysis	Type of measures	Web page	Contact	Further characterization	Previous application
MONERIS - MOdelling Nutrient Emissions into RIver Systems	MONERIS is a semi-static GIS-based emission model for point and diffuse sources of nutrients	H	Catchment	YES	YES	NA	To be defined case by by case			MONERIS calculates the emissions into surface waters via several independent pathways for separate catchments, which are topologically linked in a tree-like structure. Input data are taken from various sources (e.g. statistical year-books, emission inventories, digital maps etc). Those data are preprocessed to give specific values for every catchment.	Many basins - Danube River
MULINO - MULti-sectoral, INtegrated and Operational decision support system for sustainable use of water resource at the catchement scale	Operational tool aiming at supporting and guiding the Decision Makers in each step of the overall decision making process, from problem conceptualization to the choice of the best policy to solve it.	E	Catchment	YES	YES	OA	To be defined case by case	siti.feem.it/mulino/	carlo.giupponi @unimi.it	The tool integrates social, economic and environmental modelling techniques with GIS capabilities, a geo-referenced database and a multi-criteria approach for evaluating simulation results. The core structure of the tool is based on the Driving Forces-Pressure-State-Impact-Response.	Many basins
NetSyMod - Network analysis creative System Modelling Decision Support	Operational tool aiming at supporting and guiding the Decision Makers in each step of the overall decision making process, from problem	E	Catchment	YES	YES	OA	To be defined case by case	www.netsymod.eu/ mdss/	netsymod @feem.it	mDSS4, originally developed in the context of the project MULINO (MULti-sectoral, INtegrated and Operational Decision Support System for Sustainable Use of Water Resources at the Catchment Scale) and further developed and applied with a contribution of several other projects (including DSS-GUIDE, TRANSCAT, NOSTRUM-DSS, NEWATER and	Many basins

(*continued*)

Table 10 *Continued*

Tool name	Short description	Group	Spatial scale	Irrigation measures	Agricultural measures	Economic analysis	Type of measures	Web page	Contact	Further characterization	Previous application
	conceptualization to the choice of the best policy to solve it.									BRAHMATWIN) is a generic DSS developed to assist water authorities in the management of water resources. It can help to better understand or explain (to stakeholders) the problem at hand, to facilitate public participation required by the WFD, to take the edge of the conflict related to alternative water uses, to extend collaboration with and within different stakeholder groups.	
NIRAMS - Nitrogen Risk Assessment Model for Scotland	GIS-based model to calculate losses of N from diffuse pollution.	H	Catchment	YES	YES	NA	Fertilization schemes	www.macaulay.ac.uk/MRCS/gis/gis3_nirams.html	Ann Malcolm a.malcolm @macaulay.ac.uk	Estimates of annual residual N (by crop) are leached to surface and groundwaters by hydrological flows and routed to the stream system. The system uses nationally available land use, soils, topography and meteorology data sets and has been developed within the ArcView 3.2 Geographic Information System (GIS). The model has been delivered to the Scottish Executive Environment and Rural Affairs Department (SEERAD) and the Scottish Environmental Protection Agency (SEPA).	Scotland
NL-CAT - Nutrient Losses at catchment scale	NL-CAT is capable of simulating phosphorus and nitrogen losses in both the soil and surface water.	H	Basin/Catchment	YES	YES	NA	Crop rotation, Tillage operations, Fertilization schemes, Filter strips	www.euroharp.org/toolbox/onemodel_slide.php?mod=21	Oscar Schoumans @wur.nl	The model chain NL-CAT (Nutrient Losses at catchment scale) was built in order to evaluate the impact of different type of measures on the improvement of the surface water quality at catchment scale. The surface water modeling consist of the two important STONE components SWAP	The Netherlands

(*continued*)

Table 10 *Continued*

Tool name	Short description	Group	Spatial scale	Irrigation measures	Agricultural measures	Economic analysis	Type of measures	Web page	Contact	Further characterization	Previous application
										and ANIMO in combination with a surface water quantity and a surface water quality module. The SWAP module generates hydrological input to the ANIMO module simulates the nutrient cycle in soil and the nutrient leaching to groundwater and surface waters. Surface water discharges are simulated by the SURFACE WATER module while simulation of surface water quality processes and retention within a (large) catchment is performed by the NUSWALITE module. An erosion module P-USLE based on the modified and revised Universal Soil Loss Equations is implemented in a GIS-environment.	
PlanteInfo	PlanteInfo is a information and decision support system for farmers and agricultural advisers.	A	Farm	YES	YES	NA	Agicultural practices	www.planteinfo.dk	planteinfo@agrsci.dk	Most of the information is generated dynamically with models using frequently updated databases. A subscription system enables personalized information. For example, since the geographical location of the user's home is known after login, local weather data are used for model calculations. PlanteInfo can store previously entered information, like fields, crops and actions for the user for future use of the system. PlanteInfo also has a public version, which contains the facilities you see in the menu. Unfortunately, only a few facilities have English versions.	Denmark

(*continued*)

Table 10 *Continued*

Tool name	Short description	Group	Spatial scale	Irrigation measures	Agricultural measures	Economic analysis	Type of measures	Web page	Contact	Further characterization	Previous application
PlayAgriPoliS	A policy simulation game	E	Catchment		NO	OA	CAP Macro policy	www.iamo.de/PlayAgriPoliS/ playagripolis_start.html	Kathrin Happe, kellermann @iamo.de	It is an agent-based modeling approach to establish a link between agricultural policy reform and structural change. PlayAgriPoliS is currently under development. A beta version of 0.91 is now available for download.	Central and Eastern Europe
QUAL2K - River and Stream Water Quality Model	QUAL2K is a river and stream water quality model	H	Catchment	YES	YES	NA	Point and non-point loads and abstractions	www.epa.gov/ATHENS/ wwqtsc/html/qual2k.html	EPA U.S. Enviromental Protection Agency	QUAL2K (or Q2K) is a river and stream water quality model. The channel is well-mixed vertically and laterally. Steady state hydraulics. Non-uniform, steady flow is simulated. Diurnal water-quality kinetics. All water quality variables are simulated on a diurnal time scale. Heat and mass inputs. Point and non-point loads and abstractions are simulated. Q2K is implemented within the Microsoft Windows environment.	Many basins
REGFLUD - Management of Regional German River Catchments	The REGFLUD-project has overall objective to set-up a DSS aimed at reducing diffuse pollution in river catchments subject to economic feasibility and social acceptability	E	Farm/Catchment	YES	YES	OA	Reduction of nitrate - Institutional arrangements	www.rwi-essen.de/	dirk. huchtemann @rwi-essen. de	A model network consisting of an agricultural sector model, a water balance model, and a groundwater residence time/ denitrification model that enables consistent status-quo as well policy measures analyses. Evaluation of alternative agri-environmental measures with a benefit-cost approach based on interviews about the social acceptability of alternative	Germany

(*continued*)

Table 10 *Continued*

Tool name	Short description	Group	Spatial scale	Irrigation measures	Agricultural measures	Economic analysis	Type of measures	Web page	Contact	Further characterization	Previous application
										measures. Analysis of alternative forms of organization regarding an efficient and likewise nutrient surplus oriented water management. Summary and provision of results within an online DSS for the interested public.	
RIBASIM - River Basin Simulation Model	RIBASIM is a generic model package for analyzing the behaviour of river basins under various hydrological conditions.	H	Basin/Catchment	YES	YES	OA	Water allocation, re-use of drainage water by downstream users	www.wldelft.nl/soft/ribasim/int/index.html	ribasim.info@wldelft.nl	Windows-based software with a graphical user interface, a database, a simulation program and a tool for the analysis of results. The model package is a comprehensive and flexible tool which links the hydrological water inputs at various locations with the specific water-users in the basin. The tool describes a basin in terms of water sources and uses and to perform a simulation of the water allocation along a certain time horizon. It can be helpful in identifying possible water use conflicts among different types of uses, such as farmers or industries, in studying the sustainable development of the river basin itself and in planning the adequate measures to solve conflicts or generally improve the water resource status.	Many basins

(continued)

Table 10 *Continued*

Tool name	Short description	Group	Spatial scale	Irrigation measures	Agricultural measures	Economic analysis	Type of measures	Web page	Contact	Further characterization	Previous application
RIZA tools for groundwater management in the Netherlands	Simulation models representing system dynamics and addressing one or more issues in water policy	H	Catchment		YES	OA	Water and land management	www.rijkswaterstaat.nl/rws/riza/home/mona/download/pdf/tools.pdf	RIZA, Lelystad, The Netherlands	RIZA has developed a set of coherent models to support the policy management of the groundwater resources in the Netherlands. These models compute the hydrological effects of interventions on the saturated and unsaturated zone as well as the effects on agriculture, drinking water supply and nature.	The Netherlands
SDSS - Spatial Decision Support System	The tool assess the economic and environmental impacts from various best management practices (BMPs) in reducing sediment yield on rangeland watersheds.	E	Catchment	YES	YES	MI	Best management practices (BMPs) in reducing sediment yield on rangeland watersheds.	www.tucson.ars.ag.gov/sdss/	Yanxin Duan yduan@tucson ars.ag.gov	The models provide the watershed economic analysis, by maximizing the profit of a representative ranch assumed to cover the whole watershed with the constraints of production technology, resource, sediment control objectives and sustainable utilization. There are two major types of models, static and dynamic. Each model type supported variations in plant growth, grazing and ranch operations. Upland erosion was estimated through RUSLE2 and the sediment yield of a watershed was estimated from upland erosion and sediment delivery ratios.	USA

(*continued*)

Table 10 *Continued*

Tool name	Short description	Group	Spatial scale	Irrigation measures	Agricultural measures	Economic analysis	Type of measures	Web page	Contact	Further characterization	Previous application
SIGRIA-Information System on Water Management for Irrigation	The tool is used to plan new infrastructures; identify and solve situations of water conflict; mitigate the effects of annual droughts; develop and implement new policies.	E	Basin/Catchment	YES	YES	MI	SIGRIA is a component of information systems for irrigation managed by reclamation consortia		Guido Bonati bonati@inea.it	SIGRIA is based on GIS technologies and includes: a detailed map on land cover/use identifying all irrigated areas and crops grown per season; an irrigation suitability map; irrigation water networks, from the water source (dam river, etc.) to the farm or group of farms, including a database on the technical features of the network; extensive databases on the features of irrigation (i.e. crops. farms, costs). For certain areas a linear programming model on the optimal allocation of resources for irrigation at farm and basin level and a model on local water requirements by crops are available.	Italy
SIMIS - Scheme Irrigation Management Information System	Facilitating the management tasks of irrigation schemes.	A	Farm	YES	YES	MI	Day-to-day water management activities	www.fao.org/ag/agl/aglw/simis.stm	Land-and-Water@fao.org	This program is not limited to the water aspects but covers all the major issues of the day-to-day management activities and also includes control of maintenance, accounting, water fees and other relevant tasks. sector components; crop water requirements; irrigation plan; water delivery scheduling; accounting; water fees; control of maintenance activities; performance indicators.	Many basins

(*continued*)

Table 10 *Continued*

Tool name	Short description	Group	Spatial scale	Irrigation measures	Agricultural measures	Economic analysis	Type of measures	Web page	Contact	Further characterization	Previous application
SME - Spatial Modeling Environment	An integrated toolsuite for spatial modeling	E	Basin/Catchment	YES	YES	MI	Agicultural practices,Land management practices, Water management	www.uvm.edu/giee/SME3/	Robert Costanza rcostanz@zoo.uvm.edu	This environment allows users to develop models in a user-friendly, graphical environment, requiring very little knowledge of computers or computer programming. Automatic code generators construct spatial simulations and enable distributed processing over a network of parallel and serial computers, allowing transparent access to state-of-the-art computing facilities. The modeling environment imposes the constraints of modularity and hierarchy in program design, and supports the archiving of reusable modules in our Simulation Module Markup Language (SMML).	Many basins
SPAW (v6.02.75) - Soil-Plant-Air-Water	SPAW is a daily hydrologic budget model for agricultural fields and ponds (wetlands, lagoons, ponds and reservoirs).	H	Field/farm	YES	YES	NA	Multiple landscape and ponding variations	http://hydrolab.arsusda.gov/SPAW/Index.htm	Keith E. Saxton ksaxton@wsu.edu	The objective of the SPAW model was to understand and predict agricultural hydrology and its interactions with soils and crop production without undue burden of computation time or input details. The SPAW computer model simulates the daily hydrologic water budgets of agricultural landscapes by two connected routines, one for farm	

(continued)

Table 10 *Continued*

Tool name	Short description	Group	Spatial scale	Irrigation measures	Agricultural measures	Economic analysis	Type of measures	Web page	Contact	Further characterization	Previous application
										fields and a second for impoundments such as wetland ponds, lagoons or reservoirs. Climate, soil and vegetation data files for field and pond projects are selected from those prepared and stored with a system of interactive screens. Typical applications include analyses of crop water status, deep seepage, wetland inundation duration and frequency, lagoon designs, and water supply reservoir reliability.	
SWAT - Soil and Water Assessment Tool	SWAT is a river basin scale model developed to quantify the impact of land management practices in large, complex watersheds.	H	Basin/Catchment	YES	YES	NA	Land management practices, Water management	www.brc.tamus.edu/swat	Srini Srinivasan Agricultural Research Service of the US Department of Agriculture r-srinivasan @tamu.edu	It aims at assisting the Decision Maker in storing and managing water demand and supply information, in forecasting water demands, water availability, waste generation and water costs and in evaluating water development and management options. It is physically based and it is particularly suitable for predicting the effects of land use management, as well as climatic and vegetative changes on water, sediment, and agricultural chemical yields in large complex watersheds with varying soils, land use, and management conditions for a long time horizon (up to a hundred years). SWAT is a public domain model actively supported by the USDA Agricultural Research Service at the Grassland, Soil and Water Research Laboratory in Temple, Texas, USA.	Many basins also in Europe

(*continued*)

Table 10 *Continued*

Tool name	Short description	Group	Spatial scale	Irrigation measures	Agricultural measures	Economic analysis	Type of measures	Web page	Contact	Further characterization	Previous application
SWIM- Soil and Water Integrated Model	A simulation tools for hydrological cycle, erosion, vegetation growth and nutrient transport in mesoscale watershed	H	Basin/Catchment	YES	YES	NA	Land use change Agricultural practices Water management		Valentina Krysanova valen@pik-potsdam.de	SWIM aims to analyze climate change and land use change impacts on hydrology and water quality at the regional scale. SWIM is based on two previously developed models - SWAT (Arnold et al., 1993) and MATSALU (Krysanova et al., 1989). SWIM includes modules from both predecessors, trying to combine their advantages (hydrological submodel and GRASS interface from SWAT; spatial disaggregation scheme and nutrient modules from MATSALU), and to avoid over-parameterization. A simplified EPIC approach (Williams et al., 1984) is used for simulating all the crops and natural vegetation, using parameter values for each plant type from the database.	Germany
SWIM - Soil Water Infiltration and Movement	Tool for simulating infiltration, evapotranspiration and redistribution	H	Basin	YES	YES	NA	Agricultural practices	www.clw.csiro.au/products/swim/	CSIRO Land and Water Bag 10 Clayton South VIC 3169 Australia	The overall purpose of the model is to address issues relating to the soil water and solute balance. As such it is a research tool that can be integrated in laboratory and field studies concerned with soil water and solute transport. It is also eminently suitable for management and education.	Many basins

(*continued*)

Table 10 *Continued*

Tool name	Short description	Group	Spatial scale	Irrigation measures	Agricultural measures	Economic analysis	Type of measures	Web page	Contact	Further characterization	Previous application
TRK (N) - TRK (SOILNDB/ HBV-N)	A tool to simulate N-losses	H	Basin/Catchment	YES	YES	NA	Nutrient management and the effect on N-losses - Land use changes and water measures	euroharp.org/pd/pd/models /TRK-short.htm	Helene Ejhed, helene.ejhed@ ivl.se	The TRK system combines; 1. Preparation of areal distribution of different land-use categories and positioning of point sources using GIS; 2. Calculations of concentration and areal losses of diffuse sources (for N from arable land by using the dynamic soil profile model SOILNDB); 3. Calculations of the water balance (by using the distributed dynamic HBV model) and N transport and retention processes in water (by using the model HBV-N). The results are presented in the GIS, and source apportionment is made for each sub-basin as well as for the whole river basins. The results from the system have been used for international reports on the transport to the sea, for assessment of the reduction of the anthropogenic load on the sea and for guidance on effective measures for reducing the load on the sea on a national scale.	Nordic countries and Sweden
WATER-WARE - A Water Resources Management Information System	WaterWare is an integrated, model-based information and decision support system for water resources management.	H	Basin/Catchment	YES	YES	OA	Strategies for river and groundwater pollution-control schemes	www.ess.co.at/ WATERWARE	Environmental Software & Services GmbH, Advanced Computer Applications (ESS-ACA)	WaterWare is a comprehensive decision-support system (DSS) for integrated river basin planning capable of addressing a wide range of issues such as: determining the limits of development; evaluating the impact of new environmental legislation; deciding what, where and when new resources should be developed; assessing the environmental im	Many basins

(continued)

Table 10 *Continued*

Tool name	Short description	Group	Spatial scale	Irrigation measures	Agricultural measures	Economic analysis	Type of measures	Web page	Contact	Further characterization	Previous application
										pact water-related development; formulating strategies for river and groundwater pollution-control schemes; etc. WaterWare supports the integration of databases, GIS, simulation and optimization models, and analytical tools into a common framework. WaterWare Release 5.1 is fully web based access.	
WATSIM - World Agricultural Trade Simulation Model	CAP macro level, World Agricultural Trade Simulation Model	E	EU/State/Region	NO	NO	MA	Import/export	www.agp.uni-bonn.de/agpo/rsrch/watsim/wats_ov_e.htm	Arnim Kuhn, arnim.kuhn@ilr.uni-bonn.de	WATSIM is a recursive-dynamic, spatial world trade model for agricultural commodities. In its current version, it covers 12 regions and 29 commodities. Simulations are running from the year 2000 to 2010. Policies covered are ad-valorem and specific tariffs, tariff rate quotas (TRQs), safeguards (flexible tariffs), export subsidies, and production quotas. These policies are explicitly modelled by formulating the model as an MCP (mixed complementarity problem). The most important application of WATSIM is the medium-term analysis of trade policy changes.	Many basins

(*continued*)

Table 10 *Continued*

Tool name	Short description	Group	Spatial scale	Irrigation measures	Agricultural measures	Economic analysis	Type of measures	Web page	Contact	Further characterization	Previous application
WEAP - Water Evaluation And Planning system	WEAP is a user-friendly software tool that takes an integrated approach to water resources planning.	E	Catchment	NO	NO	OA	Alternative water development and allocation scenarios.	www.weap21.org/	Stockholm Environment Institute's Boston Center at the Tellus Institute, USA	The identification and evaluation of the impacts of climate change on water for agriculture, recreation, hydropower generation, water for municipal and industrial use, habitat function and health, biodiversity, water purification alternative strategies to study the water costs in watersheds. The tool support: the assessment of water supply augmentation through an inter-basin transfer within firm yield analysis; the resolution of water use conflicts in river basins; the development of supply and demand balances.	Many basins
WetSpa - A distributed model for Water and Energy Transfer between Soil, Plants and Atmosphere	GIS-based hydrologic model	H	Catchment	YES	NO	NA	Land use change	www.tiszariver.com	info@tiszariver.com	The WetSpa model is a GIS-based hydro-logic model, which simulates hydrological processes continuously both in space and time with the aim to simulate the hydrological behaviour of a catchment with regard to flood prediction, landuse and climate change scenario analysis and water management. The pollutant transport module aims to simulate the transport of phosphorus through a catchment. Domain covered: hydrology, landuse and climate change, water management, phosphorus transport.	Europe

(*continued*)

Table 10 *Continued*

Tool name	Short description	Group	Spatial scale	Irrigation measures	Agricultural measures	Economic analysis	Type of measures	Web page	Contact	Further characterization	Previous application
WSM - Water StrategyMan-DSS	A DSS for water planning and management of water in arid regions	E	Farm/Catchment	YES	YES	MI	Supply enhancement. Demand management. Socio-economic measures.	environ.chemeng.ntua.gr/wsm/	Dionysis Assimacopoulos assim@chemeng.ntua.gr	The GIS-based Decision Support System developed for the WaterStrategyMan Project (WSM DSS) aims to assess the state of a water resource system in terms of sources, usage, water cycles (pathways) and environmental quality. In addition, it evaluates the effects of actions, on the basis of different scenarios, alternatives and policies. Water allocation is performed according to a set of demand and supply priorities reflecting the pricing system, social preferences, environmental constraints and development priorities. The WSM DSS entails four types of management options: supply enhancement, demand management, socio-economic measures, management strategies.	Greece

Guidance Report III.2

Decision Support Systems for water resources management: Current state and guidelines for tool development

Carlo Giupponi, Jaroslaw Mysiak, Yaella Depietri and Marco Tamaro

Abstract

This guidance report reviews and summarises the most recent achievements and the current issues in the domain of Decision Support Systems (DSS) with specific reference to the field of water resources management. The focus is on the role of DSS tools for river basin management, having the EU Water Framework Directive (WFD) as the most important policy reference. The report investigates the reasons for the limited operational implementations of the DSS tools in the domain of interest, in order to learn from previous experiences and thus be able to propose a list of best practices in form of Guidelines for future developments and applications.

The targeted audience is broad and includes policy makers and scientists. To approach such a great variety of readers, the format of this report is subdivided in sections, with an executive summary presenting the contents of each section, allowing the reader to select the parts of interest. Table 2.1 provides a key for the different categories of readers, identified with the four categories of actors usually targeted by the activities of the Harmoni-CA concerted action.

This guidance report may be useful for various reasons. First of all it may be useful to assess the actual return of research efforts in the development of DSS tools, since the scientific projects designed to develop DSS are not normally required to monitor the implementation and application of the DSS beyond the end of the research grant. Another reason is to trigger a wider discussion about best practices in scientific policy support in general. Section 7 provides specific guidelines for future developments of DSS tools that could be of interest for all the four main categories of actors (in particular for the technology providers) and also for the funding agency (e.g. the E.C.) that may find useful suggestion for targeting future investigations.

The contributions of end users to the survey, and their involvement in writing and revising the document represents the main novelty of this report, as compared to several

 Decision Support for Water Framework Directive Implementation Volume 3. Edited by Peter Vanrolleghem. ISBN: 9781843393276. Published by IWA Publishing, London, UK.

Table 2.1 Readers' guide to guidance report III.2

Part of the guidance document	Main beneficiaries: Methodology providers	Technology providers	Operational managers	Policy makers
(1) Executive summary	**	**	**	***
(2) Introduction	***	***	***	***
(3) Methodological background and literature review	***	***	*	*
(4) The perspectives of DSS' end-users	*	*	***	*
(5) Review of potential end users needs that could be met by DSS tools	**	**	***	**
(6) Drivers of the implementation success or failure	***	*	*	*
(7) Guidelines for the development, implementation and application of DSS tools	**	***	**	**

other reviews and assessment efforts on the same topic recently released by research projects or published in the scientific literature.

The evidences collected during the survey and literature review were merged with the experiences of the authors in various projects, including Mulino, DSS-Guide, Transcat, Harmoni-CA, Newater, Nostrum-DSS, and Brahmatwinn.

Keywords
Decision Support Systems, Mathematical Modelling, Water Quality Management

1. EXECUTIVE SUMMARY

1.1. Water Management DSS

There are many different understandings of what Decision Support Systems are, what they are composed by, what they do (or should do) and how. They include a vast variety of methods and tools developed for diversified purposes and contexts and for that reason providing a unique definition is practically impossible. A computer-based tool is surely one important component, but more and more DSS is intended as a broader combination of the tool(s) and the process of structuring problems and aiding decisions (Section 2 provides an introduction to DSS concepts and terminology).

DSS in the water management sector usually consist of simulation models, and/or of techniques and methods for decision analysis, recently extended to include the support to participatory processes. Therefore, a DSS typically integrates multi-source geographically referenced data and data management systems, a variety of models and elaboration procedures within a customized user interface. Emphasis is given to hydrologic models accompanied by environmental assessment and/or socio-economic evaluation. The models include both those aimed at reconstructing and simulating the physical reality, and those constructed to manage divergent objectives and to find a compromise among the expectations of different actors in a participatory process.

In an idealized view DSS should act as mediators between science and policy/decision making and as catalysts of trans-disciplinary research.

Methodological proposals and tools have been developed since the 1970's. Unfortunately, DSS have found very limited implementations in the real world, thus demonstrating that most of the DSS tools developed so far have failed to meet the objective of being used in the real world (in Section 3 a brief methodological overview and a literature review are provided).

The question arises then, whether realism could allow us to make proper and effective use of – improved – DSS, or whether abandonment should be preferred. We will not provide a definitive answer, but we will identify a set of best practices to be implemented with realism, to provide the ground for more effective future developments and applications (see Box A below and Section 7 for details).

1.2. Models, DSS tools and Integrated Water Resources Management (IWRM)

As the conflicts for water have exacerbated and the policies have become more articulated and complex, more scientifically robust methods are needed by managers and policy makers. The challenges imposed by the implementation process of the Water Framework Directive (WFD) and the putting in practice of the principles of Integrated Water Resources Management (IWRM) are emblematic in this regard. Those challenges may represent an important triggering factor for the development of improved DSS tools and, indeed, the 5th and 6th Framework Programmes of European research mobilized significant financial resources, targeting that process.

Unfortunately, in the attempt to cope with issues of increasing complexity, methods and computer tools (simulation models in particular) have shown a tendency to become more and more sophisticated and complicated, and there is a growing gap between the specialized knowledge of the DSS developers and the application of this knowledge in decision making.

1.3. The perspectives of DSS end users

In order to investigate the reasons for the frequent failures of DSS tools in being adopted by the intended end users (i.e. water managers and policy makers), the report includes substantial contributions from outside the academia and from representatives of the community of DSS end users who were involved in the writing and compiling of an ad hoc questionnaire.

The scope of the questionnaire was to acquire insights into actual and potential users' needs, expectations and satisfaction, during the tools' development and application.

As a prerequisite for consideration in this report the responses to the questionnaire given by DSS users had to be distinct from those given by DSS developers. This dramatically limited the choice of tools, since it became clear very soon that DSS applications are seldom used outside/after the original development process and/ or by users different from the software developers. The objective of examining the ten tools and collecting at least two end-users' questionnaires for each tool was very challenging and was met only after great efforts and after several selected DSS were discarded (details about the results are provided in Section 4).

Neither the selection of DSS nor the information collected through the questionnaires are intended as representative samples of the universe of tools and application cases, but at least this work is probably the first attempt to involve potential end users in assessing the outcomes and the impacts of recent research efforts in developing water management DSS.

Most of the respondents reached were specialized in the field of water planning and management; just a few were experts in modelling. In general they were advisors or policy makers, although a significant portion consisted of researchers involved in DSS' applications in the context of case studies.

Most of the users reached were at their first experience with a DSS. The main motivations for that new experience were acknowledged to be a proposed partnership from research institutions and the emerging needs or inability of the previous management system to approach increasingly complex decisions. Most frequently, the main aim for developing and acquiring the DSS was to encourage and simplify stakeholders' involvement. Other frequent motivations mentioned by the end users were the need for an enhanced identification of alternative policy options and for providing transparency to previous policy decisions. The satisfaction of specific regulatory obligations and requirements, i.e. the implementation of the WFD, was also pointed out as a target of the tool application. Section 5 goes deeper in the systematic analysis of the needs and the potential role of science.

In the users' opinion, the adopted DSS is useful in a broader context, not just informing the choice of policy, but facilitating the planning process and learning among the actors, contributing to finding a compromise among different expectations and interests.

The adoption of DSS to simplify stakeholders' involvement did not necessarily make the decision making process easier. In some cases this was due to the inadequate technical features of the systems. In other cases the use of the DSS evidenced the poor quality of the input data, or failed to provide an effective user interface or to communicate hidden uncertainties. Moreover, the lack of cooperation between the partners beyond the development project was pointed out as one reason for limited use and, in some cases, even the failure of the project to deliver the tool.

1.4. Bridging the science and policy gap?

The results of the survey evidenced the role played by research in DSS implementation in the real world. One can easily say that the adoption of a DSS tool by a potential end user, who is not involved in a research activity, is, to say the least, occasional. Therefore, even assuming that more effective dialogues between the science and policy spheres provide the ground for more effective DSS tools in the future, one of the main issues for future efforts in the fields of training and capacity building is the crucial role of mediators to encourage intended end users to adopt research outputs.

Indeed, science has been increasingly called to inform environmental policy making. Appeals to provide "useful" knowledge, i.e. one with direct policy implications, is a fundamental ethical principle for scientists. At the same time however, the dominant authority of science as the most privileged source of knowledge is being increasingly challenged. Moreover, it is crucial to realize that science and policy making, despite their interdependency, are rooted in different cultures and embodied in distinct frameworks of values, incentives and concerns. These differences have frequently led to frustrating experiences at the interface between science and policy.

Section 6 analyses the known factors of DSS success or failure based on the evidence of the international literature. Despite their crucial importance, the identification of DSS success factors and their measurement is a difficult task, since the development and application of DSS entail multiple potential benefits. The unambiguous detection of DSS failure is at least as difficult as the measurement of its success. From the experiences gathered so far it appears clear in general that the process of policy making is at least as important as its outcomes. Therefore, one important criterion is the degree of change to the usual management introduced by the implementation of the DSS. This could have a dual meaning, since an acknowledged limiting factor is represented by the resistance of managers to changes in conventional practices.

Box A: List of recommendations described in Section 7

Phase A: before the development or acquisition of a DSS tool

(1) Investigate and describe the problem at hand, the resources available and the data issue
(2) Identify the actors involved in the decision and explore the social context
(3) Understand the institutional and normative context
(4) Identify possible constraints
(5) Arouse users' interest and initial commitment
(6) Identify and clearly communicate reasonable expectations
(7) Decide whether or not a DSS may be useful for the purposes of the case

Phase B: the development of the DSS tool, or acquisition/adaptation of existing ones

(1) Involve end users in the DSS development throughout the process
(2) Define a clear strategy and work programme and include quality assurance
(3) Adapt tools to the users' needs and not vice-versa
(4) Refine users' requirements during the development phase
(5) Assure flexibility
(6) Accurately manage and communicate uncertainty

Phase C: DSS implementation and application in the decision case

(1) Provide a coherent methodological framework in which the DSS tool should operate
(2) Provide effective documentation of the limitations of the results provided
(3) Provide adequate documentation and support materials all together with the DSS tool
(4) Train users
(5) Exploit the full potential of the tool in supporting the whole process
(6) Provide the basis for maintenance and further development
(7) Adopt strategies for dissemination and technology transfer

1.5. The guidelines for DSS development and implementation

The final part of the report (Section 7) includes the Guidelines for the development, implementation and application of DSS tools. That section includes a series of recommendations that have been developed to increase the probability of future successes in DSS developments for the practical implementation of the IWRM principles. They have been organised according to a generic temporal sequence, by identifying three main phases:

1. the phase prior to the actual development or acquisition of the DSS tool;
2. the phase of development (in case of new tools to be implemented) or acquisition/adaptation of existing ones;
3. the phase of implementation and application in the decision case.

The most important messages and keywords to be extracted for the Guidelines are:

- accurate preliminary exploration of the problem and identification of actors involved;
- framing of the process and the DSS tool within the existing institutional setting;
- early involvement of end-users;
- match between DSS requirements and data availability and local knowledge;
- match between local practices and the DSS procedure;
- flexibility;
- communication within and outside the group of users;
- documentation;
- training and capacity building;
- dissemination and maintenance of the tools.

Details about the results of the DSS and questionnaire surveys care reported in the Appendices.

2. INTRODUCTION

2.1. Preamble

There are many different understandings of what a DSS is, what it is composed by, what it does (or should do) and how. Many different definitions exist and we prefer to avoid proposing a new one, letting the readers develop or keep their own ideas. The concept is indeed usually related to a computer-based tool, which is surely one important component, but more and more DSS is intended as a broader combination of the tool(s) and the process of structuring problems and aiding decisions.

Decision Support Systems (DSS) pervade all domains of water management (e.g. water allocation, flood prevention and forecasting, reservoir management) for a variety of reasons. In general, we may say that researchers are attracted by the capacity of DSS to convey scientific, transdisciplinary knowledge in an accessible format to policy and decision makers.

In an idealised view DSSs act as mediators between science and policy/decision making and as catalysts of trans-disciplinary research. Again, in an ideal situation, DSS tools facilitate dialogue between policy makers and scientists and both groups benefit equally from mutual learning. In the light of increasingly complex and interconnected environmental problems, the science sphere is increasingly called to inform (and improve) actual management and planning practices and scientists are requested to produce "useful" (i.e. applicable context-sensitive) knowledge.

DSS have been developed for over three decades with the purpose of replying to those calls. Unfortunately, DSS are seldom used for the intended purpose due to the misunderstanding of the motivations and incentives leading to DSS development. Even the term DSS itself is a catchword, having different meanings for different people, and covering different interests.

Different situations can be observed in the various application fields, but as far as DSS in the water management sector is concerned, one may easily say that we have already passed the typical initial phase of euphoria and we are now quite often in the disappointment phase, which may be followed by either abandonment or realism (Matthies *et al.*, 2007). The question arises then, whether realism could allow us to make proper and effective use of – improved – DSS, or whether abandonment should be preferred. We will not provide a definitive answer, but we will identify a set of best practices to be implemented with realism, to provide the ground for more effective future developments and applications. Moreover, in order to gain more insights on that question, a questionnaire has been specifically designed and disseminated among DSS and users. The survey conducted focused on a selection of 10 recently released DSS tools and was mainly targeted to the products of European funded research. Neither the selection of DSS nor the information collected through the questionnaires are intended as representative samples of the universe of tools and application cases. The wide range of issues addressed, the diversified typologies of tools, the varied application contexts, are some of the reasons why a comprehensive and representative assessment would have required a whole multi-annual research project.

Right from the beginning of this report it is worth mentioning the difficulties encountered in identifying DSS tools used outside the development context and/or by users distinct from the software developers.

We will examine in detail the theoretical and terminological issues (see Box 1 for some details). Throughout the report the term DSS is used as a synonym for a wide range of tools (e.g. planning support systems, expert systems, etc.) which may be differentiated elsewhere, sometimes for good reasons. The issues addressed in this report are to a large extent similar for all the tools and the variety of meanings associated with the term DSS, which in this case may be seen as an advantage rather than a source of frustration.

Box 1: A bit of DSS history and terminology

The concept of DSS emerged in the 1970s when it was proposed for computerised systems providing assistance in dealing with semi-structured and unstructured problems. Gorry and Scott-Morton (1971) defined DSS as an interactive computer based system, which helps decision makers utilise data and models to solve unstructured problems. Kenn and Scott Morton (1978) proposed a similar definition, stressing the ability of DSS to couple the intellectual resources of individuals with the capabilities of the computer to improve the quality of decisions. Over the next thirty years, new issues such as the increasing severity of environmental problems and growing conflicts in the exploitation of natural resources have added new challenges to DSS development. Due to the rapid advances in computer science and related research fields, the boundaries of DSS have expanded such that the understanding of what a DSS is (or might be) has become less apparent. According to Power (1997), any system supporting decision-making, including executive information systems, executive support systems, geographic information systems, online analytical processing and software agents, may be called a DSS. More recently, emphasis was placed on the multi-perspective exploration of problems (Shim *et al.*, 2002) and on shifting the target from semi-structured or unstructured to "wicked" problems (Beynon *et al.*, 2002; Courtney, 2001; McCown, 2002; Rauscher, 1999) The ambiguity of DSS definition has been discussed by several authors (e.g. Keen, 1981), some of whom go even further by viewing DSS as having matured to the point where they have lost their identity and became part of the mainstream field of management information systems (Carlsson and Turban, 2002).

2.2 What is a DSS tool?

In Box 1 we refer to several definitions of DSS, each of them is defensible in a specific context. The most common definitions refer to DSS as a computer-based tool, a higher form of information system (Keenan, 1998). When asked to explain with their own words what they refer to as a DSS, many developers point to a set of rules, tips or methods. As stated above, currently DSSs are quite often intended as the combination of the tool(s) and the process of structuring problems and aiding decisions.

Many DSS consist of simulation model(s), and/or of techniques and methods for decision analysis. Models' roles are almost as diverse as the uses and modelling paradigms employed. The variety is comprehensible, given the critical importance of models as instruments for scientific investigation and policy making (Morrison and Morgan, 1999; Pielke, 2003). In general, models can be used to (i) measure and represent; (ii) describe structure, behaviour and pattern; (iii) reconstruct past or predict future developments; (iv) generate and test theories and hypotheses; (v) surface, encode, transfer, evaluate and interpret knowledge; (vi) guide development and assessment of policies; and (vii) facilitate collective learning and settlement of disputes (Beven, 2002; Jakeman *et al.*, 2006; Morton, 1990). Decision analysis (DA) helps to avoid biases in judgement and to make decisions more compatible with normative axioms of rationality for situations involving multiple, conflicting interests and beliefs. The trade-offs or preferences are value judgements, which are frequently not observable and must be revealed or approximated. Such uncovered preferences are context-specific and depend on the description and framing of a problem, and how the questions are formulated. For example, to assess the environmental costs of irrigation, one must consider the value of wetlands and riverine ecosystems deprived by water abstraction.

There is a broad agreement about what DSS are expected to do: in general terms, DSS are set to aid decisions and structure problems. DSS are expected to explore multiple perspectives of the problem at hand; enhance decision makers' insight into the problem drivers and policy outcomes; and facilitate communication and knowledge transfer between the actors involved in or affected by the decision. Therefore, the DSS role should be intended as tools for supporting the process of decision making much more than tools to provide answers to decision makers' questions. This is particularly true also because DA methods act as "lenses" through which the policy problem is viewed, lenses may have different curvatures and thus different DMs may (frequently do) lead to different conclusions.

The fact that alternative models or, for instance, different decision rules in multi-criteria analysis, may lead to different solutions of the problem at hand is a crucial issue for DSS tools. Adequate quality assessment procedures guarantee only to some extent that the methods and tools adopted are theoretically and methodologically robust. In some cases different approaches, all suitable for the issue in question, still provide contrasting results. This problem should not be neglected and should support the advisability of testing – as far as possible – alternative approaches in the same case.

Box 2: Glossary of terms adopted in this report

Decision: The choice of one from among a number of alternatives; a statement indicating a commitment to a specific course of action.
Decision maker: An executive person or group responsible for land-use policy, action and allocation of resources.
DSS developer: The researchers or consultants directly involved in the development of the DSS tools, for their design, coding, etc., in response to specific needs of one or more potential end user(s).
End user: Those policy-makers, consultants, technical staff or researchers, involved in DSS application, and distinct from the tool developers. The potential users, which contributed to the design and implementation of the system during the development time, were also considered.
Knowledge: This term is used to mean the confident understanding and mastery of a subject. In the present context we refer to k. as what is known by qualified individuals (expert k.), or by people in a given community (local k.). Knowledge acquisition involves complex cognitive processes: perception, learning, communication, association, and reasoning.
Model: A simplified representation of reality used to simulate a process, understand a situation, predict an outcome, or analyse a problem.
Policy vs Science: When distinction between the two spheres is made, we intend to identify within the first one the community of people belonging to private or public bodies formally sharing the burden and the power to take decisions, while in the second we include all the people (researchers, technicians, consultants, et.), who are not directly involved in the decision itself, but (may) provide technical and scientific support to it.
Policy maker: A person with power to influence or determine policies and practices at an international, national, regional, or local level.
Stakeholders/Actors: Those who have interests in a particular decision, either as individuals or as representatives of a group. Including people who can influence a decision, as well as those affected by it.

The unstructured nature of many problems related to natural resources management may be caused by: (i) different ways in which the involved persons see the problem; (ii) complexity of environmental systems, their dynamics and interconnections; (iii) identification and assessment of actions; (iv) modelling the consequences of a different course of action; or (v) ambiguous and conflicting objectives. The solution of one aspect of an unstructured problem may reveal another, more complex problem. In such situations, the process of solving a problem is the same as the process of understanding its nature; they are mutually concomitant. Only concrete solutions which have been formulated and analyzed provide a deeper insight and understanding of what the problem actually is.

Many authors consider DSS as a technology. Environmental management, as perhaps no other field, relies on a huge amount of information (e.g. GIS and remote sensing data, time series of precipitations). To store, visualize and analyze these data, the pure cognitive capability of humans is of limited use. Besides, DSS help to avoid biases in judgement and make decisions more compatible with normative axioms of rationality for situations involving multiple, conflicting interests and beliefs. To qualify for DSS, the tools are expected to turn information into knowledge. As computerized tools, DSS are surrounded by a number of issues related to technology, its diffusion and acceptance. Seeing DSS' limited success as a failure of technology tips into a number of issues including personal aspects (e.g. prior expectation, education, value and belief of intended users and impact on their job); technical aspects (e.g. user interface, performance, reliability); institutional aspects; management processes that DSS are designed to aid; or any combination of the Above. It is this viewpoint which prompts increasing frustration, negative attitudes towards DSIT and abandonment.

However, hardly any DSS research project aimed at technology in the first place. DSS were supposed to connect disciplinary knowledge and solve methodological differences, in other words, the motivation to develop DSS was to facilitate integration and transdisciplinary research. For DSS to fulfil their task, i.e. aid policy making, a number of quite distinct issues come into play. Superficially these issues may be seen as problems of data collection (different data are connected for different spatial or temporal units), the nature of data (e.g. qualitative versus quantitative), or as a lack of common terminology (e.g. a wider range of different interpretations of uncertainty). The most difficult issues to overcome, however, are problems of epistemological nature, i.e. what is recognized as knowledge, for what purpose, embedded in wider disciplinary differences. Methodological differences between social and natural sciences are notoriously known but significant differences exist even within a single discipline (e.g. ecological economics and environmental economics). The perspective of DSS as a catalyst of trans- or interdisciplinary research provides different views on the success and failure.

3. METHODOLOGICAL BACKGROUND AND LITERATURE REVIEW

Water resource management has attracted a particularly vast level of attention in the recent decades. This is partly because of the fundamental value water exhibits for sustaining life and development. Water is a finite, depletable, nonsubstitutable and unevenly distributed resource with a unidirectional flow pattern, which incites conflicts and prompts a number of threats such as floods, landslides and erosion. Most of the problems handled by water resource management have the characteristics of unstructured problems. To ensure an efficient allocation and protection of water, a holistic (integrated/comprehensive) management based on the principles of the ecosystem approach was endorsed by a broad scientific and policy community. Such a management favours pro-active, non-structural and demand-side interventions. Yet, the integrated ways to resource management require a trade-off to be made between different, often mutually exclusive interests and values, which aggravate conflicts (Walker *et al.*, 2001). As the water resource conflicts become more complex, more accurate and comprehensive evaluation of management activities is required. The process of policy making has increasingly favoured interdisciplinary, pluralistic, and inclusive methodologies, with scientists participating alongside other stakeholders in deliberative decision making, participatory assessment or group model building.

To support the implementation of the new water policy introducing the principles of an integrated, river basin management, the European Commission has dedicated a key action under the 5th Framework Programme (FP5, 1999-2001) to issues related to sustainable management and water quality. The WFD principles like transboundary management of water issues; holistic, cross-disciplinary management regimes; public involvement and application of economic instruments such as water pricing and cost recovery required further research before becoming operable in different conditions across the EU countries. The EU work programme highlighted the "social, economic and cultural pressures" on water resources, causing threats (to the environment) and increasing conflicts between water uses. It emphasised the need to improve the knowledge and technologies supporting a "rational and sustainable" management of water resources which requires policies, tools and methods able to govern interwoven environmental, social (institutional) and economic systems in their full interaction at the river basin scale.

The DSS developed in the context of EU funded research (e.g. MULINO, TRANSCAT, MERIT, GOUVERNE) facilitate WFD implementation by (i) making for effective data management, (ii) providing a suitable environment for the implementation and integration of environmental models, and (iii) aiding social learning processes among stakeholders involved or affected by the measures to achieve the WFD's objectives. Within the 5th Research Framework Programme (5FP) altogether 150 projects were funded under the water key action with the total budget amounting to 244 mil Euro. 43 projects or 68.6 mil Euro were dedicated to integrated management methodologies and tools at catchment/ river basin scale. 7 projects with a budget amounting to 11.2 mil Euro were funded under the priority 1.1.3. "Operational management schemes and decision support". Although pursuing distinctive objectives, all the projects paid attention to integrated catchment modelling, risk management, quality assessment, optimization and responses of river basins to policy changes (EC, 2001).

A review of existing DSS was carried out in a number of projects (e.g. NEWATER, AQUASTRESS, FLOODSITE, NOSTRUM-DSS, WATERSTRATEGYMAN, CODES, TRANSCAT and MULINO). The reviewed tools related to each project's specific objectives and application conditions. Altogether the review encompasses more than 150 tools and DSS application reports (see also Annex I). Also, what was considered as a tool varied considerably, including conceptual tools and process based tools facilitating learning. Most tools addressed in detail, however, were computer-based. Different evaluation criteria have been used in different projects to assess system applicability (see Box 3). Only a few reviews addressed the lack of implementation success in more detail. The variety of evaluation criteria employed suggests that the evaluation is context-dependent and experiences from one case may be only poor indicators of performance for other cases (Jeffrey *et al.*, 2005).

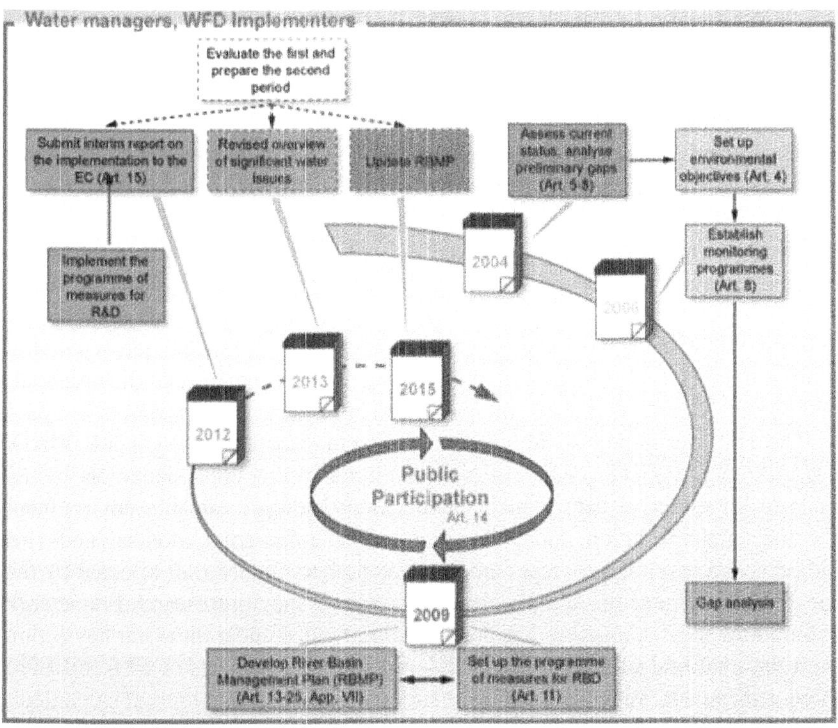

Figure 3.1 The Water Framework Directive Implementation Plan

The DSS developed to assist water resource management are mostly model-driven DSS. DSS help to understand feedback among processes operating at different spatial and temporal scales (National Research Council, 1999). Emphasis is placed on hydrologic models accompanied by environmental assessment and/or socio-economic evaluation. The application of models is referred to both the environmental models aimed at reconstructing the reality, and decision models constructed to compromise conflicting objectives and to mitigate the expectation of different actors. A DSS for water resource management typically integrates multi-source geographically referenced data and data management systems, a variety of models and optimization procedures within a customized user interface. Additionally, pre- and post-processor utilities may facilitate data exchange between the implemented models. A set of other functions may assist in data preparation and management tasks. An educational component may assist the decision maker to understand the data and implemented models. The systems may support graphical (network-based) ways to define water resource systems or are based on (or extendable by) a commercial or public domain Geographic Information System. The multi-source data bases may include environmental (hydrological, climate, soil, and meteorological) and socio-economic data (demographics, economic development etc.).

The variety of models includes hydrologic models (run-off, sediment, flood routing, flood inundation, reservoir, point and non-point pollution source models, etc.), physically-based simulation models (e.g. soil erosion, agricultural-chemical transport, crop-soil-water interaction, nutrient and pesticide transport models), socio-economic and environmental impact assessment models (e.g. crop assessment model), management and optimization models (linear programming, dynamic

Box 3: Evaluation criteria for applicability of DSS in specific contexts

The identification of DSS success factors and their measurement is not an easy task. This is because the development and application of DSS entail multiple tangible (e.g. better outcomes) and intangible benefits (e.g. mutual learning, trust and cooperation). For more detail see section 6. Here we refer to some measures and criteria of success and failure applied in the context of various international projects this review draws on.

CoDES review (urban water management) (Kapelan *et al.*, 2005)

1. Level of integration
2. Scenario modelling support
3. Policy/intervention options
4. Sustainability criteria and indicators
5. Impact assessment

6. Use of ranking/optimisation technologies
7. Risk/uncertainty modelling
8. GUI and visualisation
9. Support for different spatial/temporal scales
10. Support for group decision making
11. Input data requirements
12. Computational efficiency
13. Calibration/validation
14. How easy/difficult is it to use the tool.

WaterStrategyMan review (water management in southern European countries) (ProGea, 2004)

1. Graphical user interface, connected to a geographic information system;
2. Framework of indicators covering a range of different topics such as water quantity and quality, climatic change, demography, sustainability, cost components, and environmental water demand;
3. Ability to develop scenarios, comparable with MCA;
4. Economic analysis

AquaStress review (water stresses in Europe) (Jeffrey *et al.*, 2005)

1. Cost: purchase price, development costs, demonstrated return on investment, cost of use, training costs, licence and maintenance costs
2. Integration: level of component integration, potential for interfacing with other tools, reuse of components, platform support, interoperability across operating systems
3. Model outputs: computational accuracy and efficiency, integration of results across phenomena and domains, platform robustness, repeatability of results, easy of understanding results, verifiable audit trails, accessible peer reviewed literature for performance and case study applications
4. Flexibility and adaptability: scenario generation, exploration of intervention options, present information in multiple formats for different stakeholders
5. Ease of use: easy of links to other knowledge bases, quality of user interface, functionality, tool accessibility, different language supported

programming, multiple criteria decision making, genetic algorithm, risk assessment, etc.). The goal of the executive interface is to provide a concise and clear presentation of the data and models and should bridge the gaps in model understanding.

DSS have been developed for a variety of situation contexts which differ significantly in pursued objectives, required functionality, modelled processes and phenomena, spatial and time scale, and supported users. Generic DSS, i.e. systems built upon methodologies and models which are applicable to a range of different management situations (e.g. AQUATOOL, RIVERWARE, WATERWARE, ELBE DSS or RAISON. AQUATOOL, TWOLE and RIVERWARE), support planning and operative decision making in complex basins, including multiple reservoirs, aquifers and demand centres. WATERWARE on the other hand, addresses a range of issues like environmental impact assessment and evaluation of measures against pollution. RAISON (Booty *et al.*, 2001) represents a kind of generic software tool for rapid prototyping and practical implementation of environmental DSS. The DSS like that described in (Lilburne *et al.*, 1998) aim at problems related to the water quantity and help to improve water use efficiency. The problems addressed are typically irrigation planning problems and allocation of water resources between headwater and downstream users. Water quality issues are addressed by a range of DSS: WATERSHEDSS, MOIRA, EFDSS, BASIN, DEMDESS, DESERT, ELBE-DSS and EFDSS. The problems they deal with encompass the identification of water quality problems, integrated assessment of point and non point source pollution, selection of appropriate management practices etc. Assessment of impacts of changed land use or industrial, forestry, urban and agricultural projects on water quality and/or other environmental impacts is the goal of systems like WAMADSS, CMSS and GIBSI. Some DSS aim at more specific problems like optimal remedial strategies to restore radionuclide contaminated fresh water environments (MOIRA) or integration of a range of models of riverine ecology (EFDSS). Real-time river basin flood control and reservoir management is supported by the DSS like SDSS (Shim *et al.*, 2002) and (Stam *et al.*, 1998). FLOODSS aims at forecasting catastrophic events and at preventing and mitigating their effects on the environment. Recently, the Internet prompted development of a new typology of distributed DSS such as AQUAVOICE (Jonoski, 2002) or Transcat (Marsik *et al.*, 2006). The development of DSS is facilitated by modelling frameworks, supporting modular model developments and re-use of existing model components such as ICSM and SME described by Argent (2006).

The benefits from the DSS development are not restricted to the final system and its suitability for the supported management processes. Research challenges associated with (or intermingled with) DSS encompass (i) investigation of how innovation makes it way to practical management processes, and how institutional arrangements can help to overcome the initial resistance and maintain commitment to pursued changes; (ii) analysis of human capabilities to comprehend processes and cause-effect relations in complex worlds; and (iii) study of how technological progress and advanced computing algorithm are able to overcome the cognitive limitations. In the water resource management, DSS field is linked to integrated assessment and holistic regime to water and related land and soil resources. Consequently, the assessment of DSS' success cannot be reduced to a single measure. The conceptual diversity inherent to the DSS term and the range of benefits from the DSS development process has to be taken into account.

The computer modelling aimed at assisting water resource management is becoming more sophisticated and the gap between specialized knowledge of the DSS developers and application of this knowledge in decision making is more apparent. The water DSS development domain is challenged by the need to manage, if not to hide, the models' complexity for developing effective interfaces with the users and to make model appropriateness and decision procedures transparent to the decision maker. This is further explained in Section 6 dedicated to DSS success and failure.

4. THE PERSPECTIVES OF DSS' END-USERS

This section aims at analysing some concrete cases of DSS tools developed in support of the Water Framework Directive, by EU funded projects or National projects. A comprehensive review has been carried out to end up with approximately 10 DSS to focus on. After a preliminary categorisation of the tools aimed at selecting samples belonging to different typologies, the main criterion adopted for the final selection was their actual present or past use by potential end users. DSS developers of the selected tools were contacted to know whether end users would have been available to fill in the questionnaires (Annex III). The review scope was to acquire insights into actual and potential users' needs, expectations and satisfaction, during the tool development and application. The analysis allowed a specific investigation on successes and failures of tool application from a user's point of view. In case of tools still under development the expectation on the tool's usefulness and required functionality were assessed.

4.1 The review

A preliminary list of 60 tools (see Annex I) was developed from several sources, such as reports of EU funded projects (i.e. (FEEM, 2005; ProGea, 2004), the technical report on DSS (McKynney, 2004), the CORDIS online database (http://cordis.europa.eu/search/index.cfm). A list of 15 tools likely to be considered in the review have been identified and included in a draft list to be able to end up with a sample of about 10 tools. The preliminary contacts with DSS tools' developers allowed to identify their willingness and interest to be involved and, as previously mentioned, the probability of establishing direct contacts with end users, which was considered a pre-requisite for the inclusion in the survey.

Some tools had to be excluded from the review as they were still under development, and/or never transferred into the public domain and/or never applied. DSS tools for which the developers provided their availability and identified a list of at least two users are in the final short list.

The search for the availability of end users of the tools and the interactions with DSS developers suggested a series of revisions in the original list. TWOLE was added as a distinct product related to the MERIT DSS, while the KerDST "Deliberation Support Tool" (DST) was selected as a representative of a group of tools first developed in the GOUVERNe project. Finally, the WFD-EXPLORER was suggested as a suitable example for the products developed at Delft Hydraulics.

End users suggested by developers were contacted via e-mail for the compilation of the questionnaire reported in Annex III, designed to collect users' perspectives on the tool's development, acquisition and application.

Box 4: Main characteristics of the DSS tools examined

Tool	Domains				Sectors			Support provided				Users		Technical features		
	Surface water	Groundwater	Socio-economy	Ecology	Water management	Land	Urban	Communication	Decision analysis	Prediction of Policy Outcomes	Identification of Options	Water Authority	Stakeholders	Multilingual	Operating system	Web Based
CatchMODS	X	X		X	X	X				X		X				
Elbe-DSS	X		X	X	X	X	X	X	X	X		X	X	X	MsWin/Xp	
KerDST			X	X				X	X		X	X	X	X		
MERIT DSS	X	X		X				X	X	X	X	X	X		MsWin	
MONERIS	X	X		X	X	X				X		X			MS-xls	
Mulino DSS			X	X	X	X	X	X	X	X	X	X			MsWin 2000	
RiverWare	X	X	X	X	X			X	X	X	X	X			UNIX/Ms Win	
TwoLe	X		X	X	X				X			X				
WaterWare			X	X			X		X	X		X			MsWin	X
WFD-Explorer				X	X			X				X	X	X	MsWin Xp/Nt	
WSM DSS			X	X	X		X	X	X		X					

4.2 The selected tools

To contemplate the variety of DSS typologies, a wide range of tools' types and functions were considered. These included software libraries and tools specifically tailored for developing a DSS for IWRM (i.e. RIVERWARE), single models (i.e. MONERIS), or tools provided of several integrated models, a database and a user interface (i.e. ELBE-DSS, WATERWARE). The tools considered are presented in Box 4 and described through various features such as the domains and sectors considered, the kind of support provided, the type of intended users, and other technical characteristics.

An extended description of the selected tools is presented in Annex II. The tools' characterizations have been prepared through the available documents and investigate the primary scope and domain of the various tools development and application, some features of the technical design, the kind of support provided. Other information included concern the distribution, maintenance, users and applications, documentation and development. The descriptions have been edited by the DSS' developers.

4.3 The questionnaire

The questionnaire is divided in four different sections (see Annex III). In the first section the questions focused on the motivations and the scopes of the development, acquisition, and application of the DSS. The next section requires the specification of the role assumed by the user during the system's development and of the conditions of usage. Gathering information on the concrete experience, use, implementation and maintenance of the system was the target of the third part of the questionnaire. This part investigates in particular the impacts on the policy making process after the tool's application. The final part of the questionnaire has been dedicated to the elicitation of benefits and satisfaction resulting from the use of the system and its different components.

A first version of the questionnaire was distributed and tested at «EUROPE-INBO 2006» 4th European Conference held in Megève (France) the 20th-22nd of September 2006. The comments received from the participants on the structure and contents of the questionnaire were taken into account for its dissemination to the intended users. Thanks to the indications of the selected DSS developers, in total 40 questionnaires have been distributed to end users and 26 replies have been collected. We ascertained a willingness to cooperate in just a few of the target persons. Receiving their feedbacks was thus a demanding task. The following paragraphs are the result of the filled questionnaires' elaboration. The details of the answers given by the 26 respondents are presented in Annex IV. The answers to the free questions are faithfully reported, meaning that the language check was not carried out on this part of the text.

4.3.1 Development, acquisition and application

Most of the users reached were at their first experience with a DSS and the main motivations a for the development and acquisition of the system were acknowledged to be a proposed partnership with research institutions, but also emerging needs or inability of the previous management system to approach increasingly complex decisions. The main objectives targeted with the development and acquisition of the DSS were the opportunity to facilitate stakeholders' involvement in the decision making process, to identify alternative policy options and to gain a better understanding of the problem at hand. The fulfilment of specific regulatory obligations and requirements, i.e. the implementation of the WFD, was also pointed out as a target of the tool application.

The majority of the users contacted couldn't say whether the implementation of the DSS imposed changes to the traditional decision making process in their institution (i.e. the answer to the specific question was "Neither agree nor disagree"). This result can be explained by considering that an important part of respondents was involved in the project's case studies for tool testing. In fact, the questionnaire was rarely filled in with reference to applications for concrete decision making processes. Nevertheless, a remarkable portion of the respondents affirmed that a change occurred in the process due to the tool's adoption.

Generally, the developed DSS have been integrated with other software previously used and the integration itself was judged important for the decision to develop and acquire the system. Most of the DSS considered incorporate a GIS and seldom a database management system or a model.

4.3.2 Personal background

Most of the respondents reached are specialised in the field of water planning and management; just a few are experts in modelling. In general they are advisors or policy makers, although a significant part of the respondents are researchers or in close relation with the academic world.

The end-users' involvement concerned mainly the identification of their current and potential needs and, secondly, the test and evaluation of prototypes during various phases of the development. It emerges how end users' involvement occurs usually at the very initial phase related to the specification of the problematic and at the final phase of tool testing. Adaptation and adjustments to the user's needs through the whole process are often missing.

MONERIS and RIVERWARE are systems developed and used for at least 10 years and are generally applied on a daily basis by end-users. Systems like WATERSTRATEGYMAN have been developed more recently, i.e. 2002/2003, and are used on a continuous basis. Other DSS are used but only occasionally.

ELBE-DSS and WFD-EXPLORER are still under development, but they have already been applied in the context of pilot case studies.

Most of the users reached interact with the system directly from their PC. The respondents' opinions on the difficulty in becoming familiar with the system were discordant. Indeed, the perception of the system's complexity varied more according to the personal background and the application context, than on the type of tool. Moreover, most of the respondents don't perceive their role in the decision making approach changing because of the adoption of the DSS. This can be explained, again, by considering that most of the tools included in the survey were developed and applied by research projects.

Box 5: Evidences from the survey in brief

Quite often research projects' products are built till they reach the state of prototype and are not further implemented and delivered in a commercial setting. Insufficient cooperation between the project's partners has been singled out as the main reason of failure in the attempt to carry on the project till the acquisition and application of the tool. This is often the case of tools developed in the frame of EU projects, aiming more at research objectives and confined to the experimental phase. For this category of DSS tools it has been hard to identify and get in contact with the intended end-users.

In those cases in which end users were involved in the research projects, they often belonged to administrations in the water management sector. It appears that their involvement is problematic for various reasons: their motivation to be involved in studies marked by an investigative and exploratory intent, the difficulties in coordinating the timing of the project with the real world case timing, when decision processes are considered; the reluctance for being officially bound from the project results in future policy choices.

The lessons learned and principles tested in the environment of these projects have been further implemented by subsequent and ongoing projects. This allows eventually to characterize the subsequent framework programmes, e.g. FP4 "Technocratic DSS", FP5 "DSS tools presented to potential users", FP6 "DSS developed with potential users".

4.3.3 Implementation, use and maintenance

In the users' opinions, the adoption of the DSS was useful in a broader context, not only to provide information on the choice of a policy, but to facilitate the planning and learning processes among the actors, contributing to finding a compromise among different expectations and interests. It has been pointed out how some tools are designed to meet regulatory obligations, giving insights for the solution of the problems encountered, especially during the implementation of the WFD. Apparently, the majority of the selected tools are still maintained and further developed.

Re-use of the system did not impose great difficulties to the users, but training and assistance were considered as a fundamental prerequisite for the successful application. Similarly, another important condition was the presence of adequately trained individuals within the institution.

4.3.4 Benefits and satisfaction

Section D provided insights in the perceived benefits and satisfaction level of DSS users. The vast majority of respondents either agreed or strongly agreed that the implementation of DSS improved the pre-existing approaches (D1). The user interface and functionality of the databases were both perceived as sources of high satisfaction for the DSS used (D2) and as the most critical components (in the open questions) for an effective use and application of the system.

An incomplete database and the lack of available data, the complexity of the various modules and the perception of the model as a black box are examples of the system's limitations cited in response to D3 and D4. The adoption of too short-term time scales, in comparison with the regulatory requirements, and the non-availability of the system as open source software were also valued as a relevant limitation for the system's adoption.

When the components of the system were questioned by the stakeholders, the disputed points mainly regarded the quality of the input data, the complexity of the user interface and hidden data's uncertainties.

The analysis of responses to D4 (future recommendations) is very interesting and the development of more friendly user interfaces appears again as the hot issue for the future. The recommendations for the improvement of similar systems in the future are summarized in Box 6.

In conclusion, having remembered the difficulties encountered in identifying a sufficient number of cases of tools that were utilized by users distinct from the developers, and that most of the cases available were still to some extent academic exercises, a general optimism emerged from the end users' experiences, related to the potentialities DSS tools have to support and improve the decision making process in water resources management. Prerequisites for more effective future developments seem to be an improved user-friendliness, and, related to that, the involvement of users in all phases of the tools' development and by providing adequate training support during and beyond the end of the development phase.

Box 6: Users' recommendations for DSS development

- About the user interface: more user-friendly interface to improve handling of model inputs and outputs /make the user-interface clearer for communication with stakeholders/better interface, more flexible data base interface/implementation of GIS interfaces.
- To ensure an adequate availability of data, improvements should mainly be made at the European scale, to provide consistent, homogeneous and reliable sources of input data.
- Technical follow-up of the project has to be carried out in a more careful way.
- Stakeholders' opinions have to be taken into account
- The use of the tool for the project's purposes and further investigation allowed to understand the difficulty of adapting such a model to suit everyone's needs. In fact, it seems that each decision support system tool should, ideally, be adaptable to different scales of analysis and databases. Open access to source codes should help.
- Regular training on new features, if useful, should be planned.
- This technology is expensive to develop and maintain. Partnership between larger Institutions is often critical to fund the next generation of this type of tools.

5. REVIEW OF POTENTIAL END USERS' NEEDS THAT COULD BE MET BY DSS TOOLS

DSS are built to meet users' needs which should be, by definition, related to a specific decision issue. In a narrow view the users' needs stem from the institutional role and the responsibility assigned to the users, or better, to their institutions. For example, the role of river basin authorities may be to develop river basin plans and manage water resources accordingly, monitor water quality and quantity, release warnings in cases of expected floods or droughts etc. For example, after the devastating tsunami in 2004, the competent authorities decided to develop a tsunami forecasting system to allow early warning and the implementation of preventive measures.

In some cases developing DSS to match institutional tasks and functions can be straightforward. In other cases the available knowledge may not be sufficient. The required information may not be available at all or not in the required form (e.g. for a specific spatial or temporal unit/scale). For example, in the absence of groundwater quality data the river basin authority may enforce decisions regarding the groundwater protection or management, but these decisions may not be optimal and can be contested. In general, users' needs can be related to all impediments preventing users from making decisions based on sound knowledge and management techniques.

Sometimes the required knowledge is not available because of the lack of scientific understanding or the variability of the factors which influence the specific situations. In these cases the investment, e.g. in the monitoring networks, does not improve the knowledge on the basis of which the decision needs to be made. What is required is an improved scientific understanding of the underlying phenomena or an introduction of governance regimes which can cope with the intrinsic uncertainties and which are able to obtain a wider consent among the affected actors. Therefore, users' needs can also refer to scientific progress (e.g. interdisciplinary research about the consequences of climate change) or to the effectiveness of institutional arrangements (e.g. ability to cope with conflicts).

Science has been increasingly called to inform environmental policy making. Appeals to provide "useful" knowledge, i.e. one with direct policy implications, is a fundamental ethical principle for scientists. At the same time however, the dominant authority of science as the most privileged source of knowledge is being increasingly challenged, and a greater openness and dialogue between all knowledgeable parties is called for. It is crucial to realise that science and policy making, despite their interdependency, are rooted in different cultures and embodied in distinct frameworks of values, incentives and concerns. Frequently, these differences have led to frustrating experiences at the interface between science and policy.

Box 7: Key Issues of WFD and Research Needs

To assist in the implementation of the WFD, the Water Directors initiated a study on the Key Issues of WFD and Research Needs. The study intended to produce a list of specific topics as input for the research community. However, neither water managers nor researchers were able to provide lists of topics in sufficient detail to match them (CIS, 2005). Nonetheless the study uncovered a number of issues, some of which are relevant for the DSS field. Frequently, the required data are either not available at all, or not in the required format or aggregation level. A variety of methods and techniques, e.g. for monitoring of water quality, are applied across various countries in international river basin districts, sometimes yielding not easily comparable results. In a number of cases, adequate measures (e.g. the pesticide directive) to comply with the WFD objectives can be applied only at EU level (CIS, 2005). Fundamental gaps in the understanding of ecological processes and the human impact on such processes were highlighted. The study recommended a closer, face-to-face collaboration between the research teams and the policy makers involved in the WFD implementation.

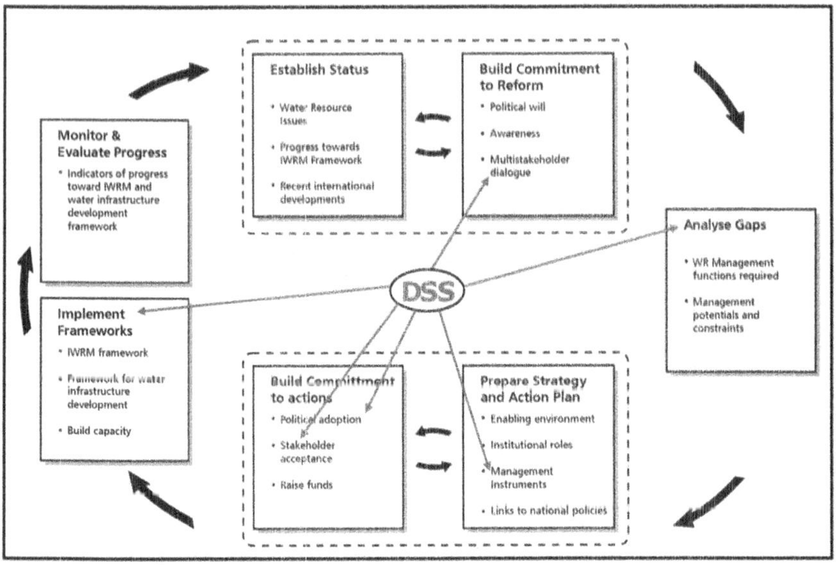

Figure 5.1 The IWRM process and the role DSS could play in supporting the implementation of the cycle. Figure modified GWP-TEC (2004)

To overcome the science – policy gap, many have argued that policy (or action) related research differs from mainstream science in several aspects: it is action-oriented (in the sense that the implementation concerns are a part of the research), integrated, value-committed (as opposed to 'value-free'), situation-specific, operating on the long term and sensitive to the lack of commitment of involved actors (Meppem, 2000; Shi, 2004). In other words the way science informs policy-making needs to be re-thought. The re-definition of relations between science and society includes release from disciplinary and institutional rigidity; methodological pluralism (embracing ambiguity); surfacing one's own normative assumptions, values, motives, potentials and limits; and an engagement in ongoing dialogue (Muller, 2003). This has several important implications: first, scientists should be prepared to facilitate policy processes rather than to determine them; and secondly, the process of policy making is at least as important as its outcomes.

A fundamental challenge of policy-related research is the quality assurance of the whole decision process and its results. Many policies are afflicted with complex conflicts and quality assurance in this context has to take into account intangible and incommensurable aspects such as subtle changes in behaviour, level of trust, and changes in relations. Their weights and how they are translated into decision outcomes (more effective and efficient decisions) and in the decision process (more informed, inclusive and transparent decision making) are to a large extent specific to a given context, and thus hardly transferable. Therefore, the peer review which is the traditional means for assessing research quality alone is not sufficient. Other criteria have been proposed, including extended peer-community review or commitment to the accountability of researchers. For example, the hurricane forecasts of the National Hurricane Center in Florida include the names of researchers who developed them. However, the lack of generally applicable quality assessment criteria poses an important challenge to the transferability of the insights collected in policy related research.

Due to the central role of policy-making models for water resource management, the quality assessment of the simulation modelling process attracted a vast attention. HarmoniQua Guidancc and Modelling Support Tool MoST (Refsgaard *et al.*, 2005; Scholten *et al.*, 2007; Scholten and Udink ten Cate, 1999) guides model development, monitors the progress done, and advises the modelling team on the basis of model journals, collected during the previous projects. To this end the modelling process has been broken down into five generic steps and forty-four detailed tasks, each further specified in terms of activities to be performed. The steps, tasks and activities have been completed with practical recommendations, further differentiated according to modelling domain, users' level of expertise, and job complexity.

6. DRIVERS OF THE IMPLEMENTATION SUCCESS OR FAILURE

6.1 Defining success

Despite their crucial importance, the identification of DSS success factors and their measurement is a difficult task. The development and application of DSS entail multiple benefits which make measurement of DSS success difficult. Many factors are likely to influence DSS success, but their importance may vary significantly according to situational variables. Very frequently studied factors of DSS success include (i) actual use and re-use of DSS by intended users, and (ii) user (perceived) satisfaction. The former indicates that the benefits from DSS usage were considered to be greater than the costs, especially in case of re-use. The latter expresses the level of the satisfaction perceived by the users as a consequence of having implemented the system. Obviously, these factors are not independent of one another, as a higher satisfaction can lead to its repetitive use and vice versa, see in (Finlay and Forghani, 1998; Wierenga and Oude Ophuis, 1997). Neither of the two factors analyses the changes in decision maker's performance achieved by using DSS, very difficult to assess on its own. The change itself should not be considered a priori as a positive indicator, first of all because change does not necessarily mean improvement and, secondly, because one recognised characteristic for the successful implementation of DSS tools is their ability to adapt to existing management approaches. Efforts required for changing usual management practices by the end users have been often identified as one of the main barriers to the uptake of the available DSS.

The DSS usage as expression of DSS success, being a twofold variable (is used, is not used) is criticised for its lack to further differentiate the systems used (Gelderman, 2002).

The pure subjective assessment of the DSS acceptance is also not unambiguous. Frequent usage of a system, familiarity with its structure and involvement in DSS building have been proven to increase the decision maker's illusion of control (Kahai *et al.*, 1998; Matthies *et al.*, 2007).

The unambiguous detection of DSS failure is at least as difficult as the measurement of its success. In many cases, even when DSS development was stopped or the system was not used by its intended users, essential learning processes took place. Newman (1999) refers to a case when the DSS has become obsolete because the end user has learned (by DSS using) its logic and was able to apply it on his own. Bell *et al.* (2001) refer to a similar situation when the decision makers, by using the DSS, were able to get detailed insight into the decision problem and understood their own preferences better, although the adopted solution differed from the solution proposed by the DSS. The examples show that if a DSS is not practically used or the solutions proposed by the systems are not adopted, it cannot be generally considered as a non-ambiguous criterion of the systems' success or failure.

6.2 Critical success factors

There is a large body of literature dedicated to the processes behind the acceptance of and resistance to DSS and technology in general, a selection of relevant evidences and references are presented below.

A number of typologies of critical success factors have been developed in management and computer sciences (Alter, 2004; Borenstein, 1998; Chen and Lee, 2003; Eicher, 1999; Finlay and Forghani, 1998; Gelderman, 2002; Harrison and Pelletier, 2001; Jiang *et al.*, 2000; Kahai *et al.*, 1998; Kamp, 1999; McBride, 1997; McHaney and Cronan, 2000; Newman, 1999; Nightingale, 2004; Palvia, 1997; Palvia and Palvia, 1999; Palvia and Chervany, 1995; Poon and Wagner, 2001; Santhanam *et al.*, 2000; Schultze and Boland Jr, 2000; Shaw and Edwards, 2005; Uran and Janssen, 2003; Wierenga and Oude Ophuis, 1997). These typologies distinguish between personal factors (such as prior expectation, education, value and belief, impact on user's job); technical aspects (user interface, performance, reliability) and the interaction of both (social content gain and loss before and after the DSS implementation), differing in the importance attached to these categories. Other categories frequently applied to study the DSS success include the external environment, the organizational aspects (e.g. changes in interpersonal relations) and the management processes that DSS are designed to aid. The role of some other actors has been reported as crucial for the DSS success (Finlay and Forghani, 1998): (i) DSS champions - managers who actively promote the DSS concept among the potential users; (ii) supporting political actors - people whose influence is needed to secure support for the DSS; and (iii) direct supervisor of the user - persons to whom the user is directly accountable. It is also practical to distinguish factors which raise commitment to a change (e.g. availability of technical and human resources, presence of a powerful actor inclined to support innovation) from those which maintain the commitment (e.g. early user's involvement, DSS prototyping).

Another class of approaches analyses cognitive aspects (cognitive styles of decision-makers) and their influence in accepting/resisting formal decision making and employed technology (Lu *et al.*, 2001; Chen and Lee, 2003; Barr and Sharda, 1997; Jiang *et al.*, 2000; Kahai *et al.*, 1998). Cognitive types are chronic motivations determining the initialization, course and cessation of information seeking and processing (Blais *et al.*, 2005). Cognition may influence perceived facts and behaviour tendencies (Lu *et al.*, 2001). Consequently, different people with the same belief may develop different attitudes and different people with the same attitude may develop different behavioural intentions. Different cognitive styles have been described to explain these phenomena. Blais *et al.* (2005) employed three cognitive style measures – the need for cognition; the personal need for structure; and the personal fear of invalidity. (Lu *et al.*, 2001) analyzed cognitive styles determined along information gathering (perception styles) and information evaluation (judgement styles), resulting in sensing/intuition and thinking/feeling cognitive styles. The emphasis on the cognitive modelling characterizes a new trend in DSS development. The cognitive DSS focus explicitly on decision maker's mental models, facilitating their enrichment, validation and supporting backward and forward thinking (Chen and Lee, 2003). Other phenomena analyzed in the context of DSS/technology acceptance include illusion of control (Kahai *et al.*, 1998), user resistance (Jiang *et al.*, 2000) and reliance effect (Barr and Sharda, 1997).

Many attempts have been made to explain/predict users' attitude towards technology (Colvin and Goh, 2005; Davis and Venkatesh, 1996; Garrity *et al.*, 2005; Hung and Chang, 2005; Karppinen, 2005; Kukafka *et al.*, 2003; Lee, 2004; Legris *et al.*, 2003; Lim, 2003; Lowry, 2002; McFarland and Hamilton, 2006; Purohit and Kandpal, 2005; Sohn and Ahn, 2003; Straub *et al.*, 1997; Wu *et al.*, 2007; Yang and Yoo, 2004). The Technology Acceptance Model TAM (Davis, 1989), based on the theory of reasoned action TRA (Fishbein and Ajzen, 1975) tried to explained the impacts of external variables on internal beliefs, attitudes and intentions. Two factors were found to determine system use: perceived ease of use (PEOU) and perceived usefulness (PU). External factors act through these beliefs. Subsequently, the TRA and TAM evolved in a number of different models such as Theory of planned behaviour TPB (Ajzen, 1985; Azjen, 1991) and decomposed TPB (Taylor and Todd, 1995). A different approach in behavioural research was taken by diffusion theory (Rogers, 1995), considering technology success as a social process related to the spread and adaptation of innovation, see also (Lee, 2004; Sohn and Ahn, 2003).

7. GUIDELINES FOR THE DEVELOPMENT, IMPLEMENTATION AND APPLICATION OF DSS TOOLS

In this section we translate the outcomes of this study into concrete recommendations for the development, implementation and application of DSS tools. Both the literature review and the survey of DSS tools and end-users opinions were framed within the context of water resources management. Nevertheless we believe that the evidences of this study may be of interest for the broader field of DSS in general.

The development and application of DSS require passing through a series of different choices. We preferred to organize these Guidelines around the idea of a generic temporal sequence, by identifying three main phases:

- Phase A: prior to the actual development or acquisition of the DSS tool;
- Phase B: of development (in case of new tools to be implemented) or acquisition/adaptation of existing ones;
- Phase C: of implementation and application in the decision case.

The sequence is not necessarily universally valid, in particular for what concerns the organisation of the work within the three main phases, because some aspects could be dealt with in parallel, for instance, or be approached at different times in the variety of real world implementations. Subsequent steps with internal iterations are possible, with "control phases" for assessing the quality of the process and its fitting to the application case.

In general, just like a check list, or a collection of recommendations, we believe that professionals and managers involved in cases in which the use of DSS tools is considered, should go through the whole list, in order to, at least, avoid common problems and mistakes that contributed to the –unfortunately very long– list of failures when approaching the development of new tools, or the practical implementations of one of the numerous tools developed in the WRM field over the last 30 years.

It is important to note that the development of a DSS is always a process involving different groups of people, playing different roles. A task force could be identified as that group of few people initiating and managing the process. It usually includes the manager(s) and/or policy/decision makers, the DSS developers, the technical staff at the competent institution, external consultants, experts of different disciplines, e.g. a sociologist for managing participatory approaches, a modeller for integrating simulation capabilities in the DSS package, etc. Other groups may include various categories of actors (stakeholders) involved in various ways and possibly at different stages in the decision process.

7.1 Phase A: before the development or acquisition of a DSS tool

(1) Investigate and describe the problem at hand, the resources available and the data issue

The nature of problems to be solved should be investigated at the beginning of any decision process. In the specific context of examining the adoption of a DSS tool, this should be done with a specific emphasis on clarifying whether or not sufficient human and financial resources could be available. Financial resources could also be needed, for example to consolidate the available data sets. Indeed, the data issue (i.e. the availability of data for informing the decision process) should be approached in order to acquire preliminary information of fundamental importance to choose among the DSS already available or the requirements for a new DSS tool to be developed. Given the complexity of the problems at hand and their specificity to local situations, expert knowledge may provide fundamental support for compensating the lack of quantitative data. In that case adequate techniques are needed for scientifically robust expert elicitation and integration in the decision process.

The outcomes of the survey with DSS users evidenced that quite often a large gap exists between DSS data requirements and data available to users. Moreover human resources (i.e. trained people available within the management administration) are also often limited. When internal human resources are not available, mediators of DSS development and implementation (e.g. external consultants) can play a crucial role at the interface between DSS developers and managers (final users), by providing the administration with the specific skills required and by filling – temporary – gaps of the internal staff.

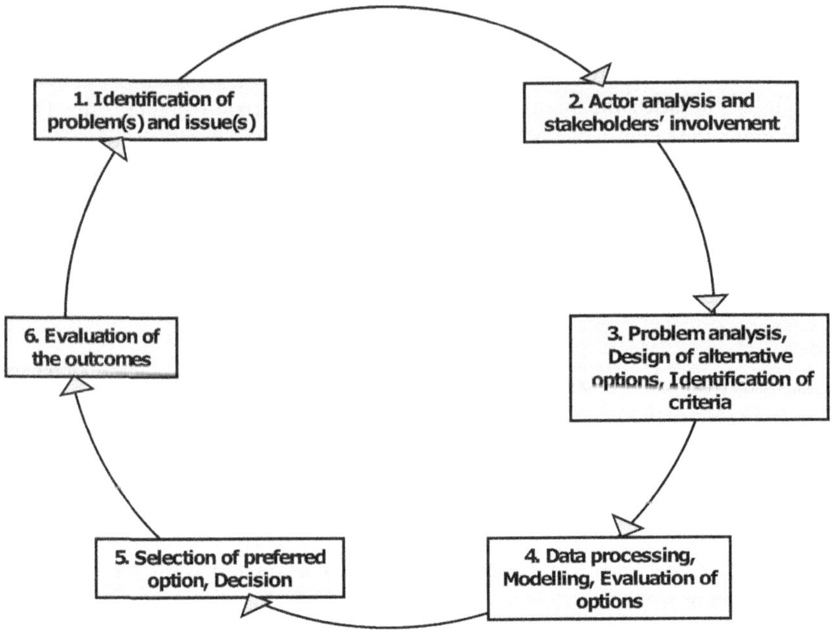

Figure 7.1 A general cycle for the implementation of a decision making process.

(2) Identify the actors involved in the decision and explore the social context

To derive a robust understanding of the problem within its social context the users together with the DSS developer(s) need first to find out whose perspectives/ viewpoints are relevant and to apply methods and techniques to make the perspectives explicit. For example:

- identification of relevant actors or categories of actors;
- selection of a representative group of actors without compromising the diversity of knowledge, values and viewpoints associated with the problems;
- exploration of the diversity of views by involved actors.

The HarmoniCop HandBook may provide more details (Ridder *et al.*, 2005).

The participatory approach is becoming a prerequisite of every legislation dealing with environmental management. DSS should be developed taking this need into account: they should be designed to allow multiple users, facilitate the integration of different disciplines and viewpoints, facilitate the transparent processing and inclusion in the DM process of data coming from PP activities.

(3) Understand the institutional and normative context

The institutional mandate of the user (what decisions users can and have to make, and on the basis of which information), the internal rules and practices within the organisation, and the relation to other institutions should be carefully investigated and clarified.

DSS may impose some degree of change to the decision-making process. It does so with the aim to improve the quality of this process, and subsequently, the decision outcomes. However, any change, be it positive or not, usually encounters a resistance of at least a part of those who are affected, thus increasing the risk of implementation failure. These risks may be reduced by actions targeted to provide timely and effective communication, including the anticipation of possible changes to be expected from the DSS implementation.

DSS impose some degree of change to the end users' practices: some sort of resistance should therefore be expected. Many known failures of DSS tools resulted from the excessive demand of change imposed to the users. Different is the case of changes imposed by new regulations; in that case DSS may be perceived to play a role in alleviating new administrative burdens. The WFD could be an important triggering factor for the adoption of new technologies, such as DSS.

(4) Identify possible constraints

The development of a DSS tool specific for the application in a decision case should adequately consider what the constraints could be to its practical implementation. Two aspects are worth mentioning:

- the technical limitations to the use by the intended users. For instance, in case of web based tools, for participative approaches, requiring efficient Internet connections.
- the limitations coming from the DSS interface and documentation, which should be carefully targeted to the intended users.

DSS interfaces are often developed for experienced users. One common problem is also the language: most DSS are developed in English, but not all the users can understand it.

(5) Raise users' interest and initial commitment

The initial users' commitment requires a clear and comprehensible specification of the system that includes what the system should provide and how. The rising of initial commitment depends upon the extent to which the users perceive the practical utility of the system for the tasks they have to carry out. For the success of DSS the support of a sponsor from within the organisation is most beneficial, preferably among the senior managers, who realise the benefits from the DSS and help to explain it to others.

Frequently quoted indicators for assessing the usefulness are:

- Perceived Usefulness (PU) is one of the central variables in models set to explain success of technology. PU is the degree to which users believe that using a particular system would enhance their job performance.
- Perceived Ease of Use (PEU) refers to beliefs that using a particular system would be free of effort.

Once the decision about the adoption of a DSS tool is taken, efforts should be targeted to raise and maintain the motivation of the whole staff of people involved in the process. This is very important in those cases in which the approach is new and comes from outside the user administration.

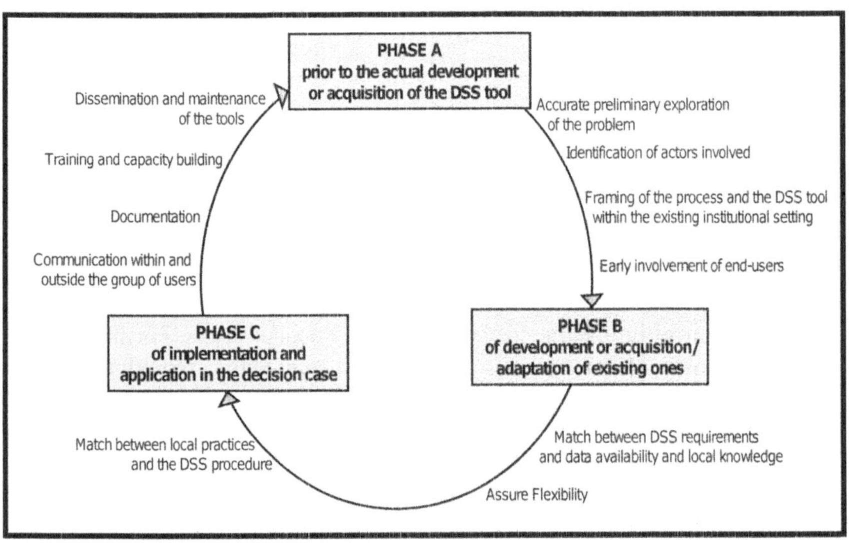

Figure 7.2 The guidelines' flowchart

The perception of ownership of the tool and its implementation by the staff of the user administration is very important for effective application and uptake in normal practices.

(6) Identify and clearly communicate reasonable expectations

DSS can assist the users in a number of ways:

- to surface various perspectives on the problem,
- to reconstruct cause-effect relations,
- to guide the development of policies or predict their outcomes,
- to facilitate collective learning and settlement of disputes,
- to improve social acceptability
- to augment or inform judgements of users and help to avoid systematic biases or inconsistent choices.

The extent to which these contributions to a specific decision process could be provided depends upon many factors, not last the availability of information. A preliminary assessment should make reasonable expectations clear to the end users.

The DSS do not stop users from making judgements, nor should they attempt to do that. Decision makers willing to delegate decisions (and power) to ICT tools are, to say the least, rare.

(7) Decide whether or not a DSS may be useful for the purposes of the case

Phase A should lead the task force to understand clearly whether or not a DSS could be useful for the decision case. Suitable tools are not always necessarily available nor are resources always sufficient to be adapted to the specific case or even for the development of a new tool from scratch (see Box 8 for more insight about this choice).

Box 8: About the decision on whether to apply an existing DSS or whether to develop a DSS from scratch

If an existing DSS has to be applied in a context for which it was not originally developed, then a transfer of knowledge and experiences takes place which may not be an easy task. The transferability is easier if the encountered problems are similar to those for which the DSS was originally developed. Generic interfaces and interoperability increases the technical transferability of DSS.

A number of web portals (such as WISE-RTD or NetSyMod) provide an overview of the available tools and help to choose one of them, according to a number of criteria. The application of existing tools is also favorable (advisable?) because it allows to re-use available tools and to further exploit previous research efforts.

The tools which have been used in practical policy making for some time are more mature, robust and less susceptible to errors or crash. In addition, users can often fall back on substantial experiences collected by others. These experiences, however, are normally not explained and form the users' tacit knowledge.

When applying an existing system, one should carefully verify the DSS' features:

- interoperability: the ability of the DSS to work together with other systems utilized in the users' institution, in order to accomplish a common task;
- data requirements: they impose costs of using the systems, and these costs need to be seen in relation with decisions which are informed by the system. Systems developed in an academic environment are more sophisticated and tend to have higher data requirements;
- methodological background: different tools may yield different results, because they may adopt different methodologies providing "different yet equally legitimate scientific lenses (assumption and theoretical underpinning) for understanding and interpreting nature" (Sarewitz and Pielke, 2000).
- built-in rationality of the tools: this refers to the way decision making is organized. The prevailing approach of current policies is to adopt the public participation (PP) paradigm, but this is not necessarily the approach adopted neither by the tools nor by the current practices of the decision/policy makers.

The development of a new DSS is a better choice in situations when the adaptation of an existing DSS is either too demanding or hardly possible. Moreover, the development process allows better involvement of users in specification and early testing of tools, which has a favorable effect on their commitment to apply the tool once available and trust in the results yielded by the tool. These benefits may under certain circumstances outweigh the higher costs and longer delivery time associated with the development of new systems.

The development of new DSS, while taking advantage of the learning process between users and developers, is facilitated by tools such as modelling

frameworks and DSS generators. These tools allow to design and deploy software 'components' which can be re-used and re-assembled according to users' requirements.
When developing a new DSS remember that:

- DSS are normally built for situations which require complex (social) choices and involve problems which are difficult to grasp in their complexity. In such situations it is difficult to obtain a shared understanding of what the problem is or how it should be tackled. Each problem is specific in its context;
- development of a DSS makes sense in particular when the developers (and the users) expect to face similar problems several times in the near future, or when the development of the system itself helps to take off the edge of the conflict.

7.2 Phase B: the development of the DSS tool, or acquisition adaptation of existing ones

(8) Involve end users throughout the process of the DSS development

Early involvement of the intended users is one of the most critical success factors, but only if both the DSS development and the decision process are ready to adapt to their needs.

As specified later, the development of a new DSS or the adaptation of an existing one is a process of mutual learning between the developers and the users. The users should not be asked to specify their needs only at the beginning of the development, but they should be consulted and involved in the system testing throughout the development process. This has several advantages: the user gains better understanding of the components constituting the system and can consequently better assess the suitability of the system for specific tasks or contexts;

- the user gains early experiences which are valuable for later practical application of the systems.
- the users feel more involved and "powerful" within the process

And disadvantages:

- the time required could constrain the availability of end users
- early involvement includes also a risk of losing critical attitude towards the system which is exemplified in overconfidence with the systems' capabilities or results.

DSS end users are often busy managers: their involvement effort should be carefully managed.

(9) Define a clear strategy and work programme and include quality assurance

There are a number of development strategies or techniques approximating the way systems evolve towards more complex techniques and functionality:

- in the waterfall method development is seen as flowing steadily downwards (like a waterfall) through the phases of requirements analysis, design, implementation, testing (validation), integration, and maintenance;
- a more advanced or flexible process of software development is the incremental and iterative system development, allowing the developer to take advantage of what was being learned during the development of earlier, incremental, deliverable versions of the system. Key steps in the process were to start with a simple implementation of a subset of the software requirements and iteratively enhance the evolving sequence of versions until the full system is implemented. At each iteration, design modifications are made and new functional capabilities are added.
- Prototyping is the process of quickly putting together a working model (a prototype) in order to test various aspects of a design, illustrate ideas or features and gather early user feedback. The availability of an initial system to react to or to improve upon may overcome users' initial difficulties to express what they expect from the system. The prototypes elicit the future expectations of users and provide the system's developers with a continuous flow of user feedback. In this model, only a low level of effort is necessary to rebuild the system after the direction of development has changed.

Quality assurance techniques may significantly contribute towards enhancing the credibility of ICT tools developed at the interfaces between science and policy. The HarmoniQuA web site (http://harmoniqua.wau.nl/) on quality assurance for catchment and river basin modelling can provide practical materials which can be useful also for DSS development in water management.

(10) Adapt tools to the needs and not vice-versa

Building lasting partnerships with the research teams that developed the system makes it easier to adapt the system to changing conditions. This must be an iterative process. Adaptation is easier if one is supported by expert knowledge (e.g. gained through active involvement in the system's development). Quite often pre-existing tools have been proposed to potential end users without an adequate adaptation to the specific needs.

DSS tools should facilitate the work of end users: the effectiveness may increase if they are adapted to build upon the existing data bases and provide outputs specifically designed to meet the formal requirements of the administration.

(11) Refine users' requirements during the development phase

As the development advances, needs, constraints, etc. can be specified in more depth. These specifications include:

- refinements on data requirements and updating available data;
- integration between the tasks fulfilled by the system with other systems used in the institutions;
- new functionalities that may be required;

Since it is often difficult for various reasons to specify the requirements on DSS in depth at the beginning, it is recommended to design the development process with enough flexibility (see next paragraph) to accommodate required reasonable changes. For an example of requirement analysis the reader is referred to the Harmon-IT report (Gregersen *et al.*, 2002).

Users' requirements may include the translation of the interface in the local language (see also recommendations 4 & 16).

(12) Assure flexibility

As the DSS development process progresses, the changes to system specification become increasingly costly and require more time. Flexibility can be enhanced e.g. by taking into account up-to-date technical standards for data description (metadata), data exchange (e.g. exchange of spatial data – OPENGIS standards), system description (UML) or component technology. Designing open source systems or providing the systems at no or low costs increases the number of developers and users.

For one of these examples the reader may refer to the Harmoni-IT report on OpenMI (Moore *et al.*, 2005).

DSS tools should be adapted to the availability of data (see also recommendation 1), but also to cope with unexpected lack of information. In those cases the ability to implement expert knowledge in the process is of fundamental importance.

(13) Accurately manage and communicate uncertainty

Uncertainty pervades all aspects of environmental policy making. One reason is that policy outcomes are only partly predictable and their associated uncertainties are large enough to sustain persistent conflicts and indecision. Related to this is the tendency for scientists to conceal uncertainty for fear of diminishing their professional credibility and encouraging indecision (Bradshaw and Borchers, 2000). It is also because:

- uncertainty provides a political resource, which can sustain personal beliefs and self-interest (Stirling, 2006; Weiss, 2002).
- Uncertainty poses various philosophical challenges regarding the origin, nature and value of knowledge, ethical challenges regarding acceptable levels of knowledge and risk, its distribution, and who has the mandate to decide, and political challenges regarding how to act when faced with substantial uncertainty.

It also poses several practical challenges, in terms of identifying and describing (quantifying, qualifying) uncertainties, propagating them through decisions and communicating the results of an uncertainty analysis.

The guidance report on uncertainty analysis can provide practical materials for further reading (Refsgaard *et al.*, 2010)

The adoption of the DSS should encourage the competent administration to deal with the various sources of uncertainty and include their treatment in the communication of results.

7.3 Phase C: DSS implementation and application in the decision case

(14) Provide a coherent methodological framework in which the DSS tool should operate

The tools per se do not guarantee the quality of the process. DSS should be framed within methodological frameworks in which all the phases and components of the policy/decision making process are considered.

The recent research projects, in particular those supported within the EU Research Frameworks have delivered a variety of methodological proposals in this field.

Various attempts have been made with the aim of facilitating the dissemination of tools (see for instance the WISE-RTD portal at http://www.wise-rtd.info/) but also of the methodological frameworks: the NetSyMoD portal is an example (http://www.netsymod.eu/), and the Guidance for Integrated Deliberative

Decision Processes developed by the Project Advisor (http://ecoman.dcea.fct.unl.pt/projects/advisor/iddp/).

(15) Provide effective documentation of the limitations of the results provided

DSS, like simulation models, encode knowledge, but they do so selectively, leaving out what is not perceived as relevant and what cannot be formalised. Preserving the relevant features, while discarding the unnecessary complexity can be the main strength of models and DSS, but it should be carefully managed. The choice of what is relevant and how this choice is made is crucial.

The communication of the process and its outcomes is always a very challenging issue in which the subjectivity and bias of the users and developers may play a relevant role.

The quality of the DSS results is determined by the quality of the decision processes and the usefulness of the outcomes depends on the quality of the communication (and training) strategy.

The results of DSS applications should be adequately documented, and assumptions, subjective choices and uncertainties of various kinds should be transparently communicated with charts, tables, and statistical annexes. Such documentation should allow interested people to go deeper step-by-step in the understanding of all the details of the decision process.

(16) Provide adequate documentation and support materials all together with the DSS tool

The final delivery package should include a handbook, methodological guide, tutorial and other additional features to increase the prospects of the system and reduce unintended misuse of the system. The inclusion of application examples can be very helpful in this regard. It is also recommended to provide the opportunities for collecting feedback and assessing the satisfaction with the tool.

An online discussion forum where the users can post their questions and report about problems/issues they experienced can be a valuable support for both the developers and users (several examples could be mentioned in particular in cases of fora/lists connecting the user's communities, for example in the case of RiverWare).

(17) Train users

Through training, the familiarity with the system increases as does the perceived usefulness of the system. Training should not be restricted to explaining how to use the software, explanations of the theoretical background and methodological strengths and weaknesses are vital for correct usage of the system. Clarifications of what the DSS cannot provide can also be helpful.

Training of users may become a crucial aspect to allow the reuse of tools outside and after the initial development environment.

(18) Exploit the full potential of the tool in supporting the whole process

The most valuable result of a DSS is the understanding of the problem at hand in its complexity. Even if the DSS yields practical recommendations such as ranking of various policies, these should not be adopted uncritically.

The success of the DSS implementation and application does not depend only upon the adoption of the results. The improvement of the quality of the decision process is the main indicator of success

Even the case of a DSS becoming obsolete because the end user has learned (by using) its logic and is able to apply it on his own (without the tool), could be considered a successful case.

(19) Provide the basis for maintenance and further development

A DSS is hardly ever finished. In order to remain useful, it has to be continuously updated. Maintenance means the commitment of the developer to assist users beyond the development project and after the final version of DSS has been released. This requirement can hardly be satisfied with the usual structure of research grants which are limited in time to 3-4 years and do not provide specific opportunities for follow up activities, such as the maintenance and dissemination of the products, training activities, etc.

Many DSS developed in an academic environment lack long-term commitment and consequently, their usability decreases in time. Development of lasting partnerships is beneficial for both developers and users.

Various strategies are possible, at the EU level, in order to provide opportunities for longer time development and maintenance:

- include requirements for dissemination and exploitation activities within the call for proposals of research projects in this field;

- facilitate transfer of knowledge to non-academic institutions – involved in project consortia – that may be interested in commercial exploitations;
- provide specific forms of financial support for facilitating the adoption of selected tools, or
- identify lists of operational tools to be used for coping with the various requirements of directives and regulations.

(20) Adopt strategies for dissemination and technology transfer

It is clear that the current system of research grants is not designed for supporting the transfer of technology from the academic environment to the policy sphere. The end of the research grant often coincides with the release of the prototype. A follow-up phase would be necessary for implementing actions targeted to the release of the operational tool and its dissemination to potential users.

DSS end users identified the need for proposals for technology transfer and for supporting the development from the prototypes to the operational tools. Moreover specific dissemination efforts are required (e.g. "a travelling team of DSS experts who can visit staff on the field and work out case-studies together.

Box 9: Comments to the Guidelines from the user perspective (Marco Tamaro)

Phase A

One of the main limitations towards the widespread application of DSS lies on the inadequate manager's perception of the role of these tools in supporting the decision making process. In fact, DSS aim more at providing complementary rather then substitutive functions.

The contributions of experts who are adequately trained for modelling of complex scenarios is fundamental. Similarly the tools which are meant to be applied by the users themselves need to be simply tailored to support specific tasks in the decision process. Point 1 in the guidelines ("Investigate and describe the problem at hand, the resources available and the data issue") is the most difficult to put in practice. All DSS should consist of suitable methodologies to allow the implementation of the process in a participatory way. It is fundamental that the process remain flexible enough to allow the adoption of the tool in more than one decision context. The use of the system for the evaluation, documentation or validation of previously adopted policies should not be considered improper, even if the full potential of the tool is not fully exploited. This type of application fosters the feedback on already implemented policies and it offers a training opportunity, useful for future applications. The test of previously taken decisions could encourage the decision maker to overcome misconceptions towards DSS. The tool needs to be adapted to the management and not vice versa. It is important to keep the data requirements as low as possible. The quality of the system's results depends more on the choice of parameters and weights, followed by a sensitivity analysis, than on the degree of the system's sophistication. Points 6 and 7 in the guidelines are very important and depend on the capacity of the consultant to communicate the usefulness and potential of the tool.

Phase B

The developer has to be part of the project's team and has to collaborate with the manager to develop a system suited to his needs. At this point brainstorming techniques, if effectively used, can reduce the time generally requested for the system's implementation. Flexibility and efficiency (i.e. it is not necessary to spend resources for the automatic import of data) are central for the system development. It is advisable to use and to import data (with cut-and-paste) from external standard applications which are used as pre-processing tools.

Uncertainty should be appreciated, in that it is naturally part of the decision-making process.

Phase C

The application of the methodology has to be coherent with the previous development phases, of course if it's not aiming at the evaluation of a decision that has been already taken.

If trained users or experts are not available in the end users' institution, the direct and autonomous use of the tool should be confined to the final phase, in which the weights and value judgements are surfaced. The tool should assist the end user by assessing the coherence of the judgements. This final phase permits the manager to gain control and prevents failures due to the managers' perception of being restricted in their decision functions. For this reason it is necessary to critically review the results (i.e. what would have happened if parameter X had been judged as more important with respect to Y or Z?).

It is the role of the developer/adviser to assure an efficient and periodical contact with the end user to elicit feedback and adapt the tool to the end users' needs.

8. OUTLOOK

This work contains many suggestions and practical advice that can be useful for future DSS developments, but it clearly does not provide a detailed recipe for success. Although many ingredients of such a recipe are known (e.g. early end-users' involvement or system flexibility), their exact dosage (quantity and quality) is to a large extent context-dependent. What has worked out in one context is no definite guarantee for success in a different context.

Nevertheless, we believe that the present work may contribute to fill a specific gap of knowledge on recent experiences in DSS developments and applications in the water management field, leading to the definition of a set of guidelines that can be considered valid in general, and helping to avoid past mistakes, knowing very well that they should be significantly tailored and adapted case by case. Therefore, the recommendations reported in Section 7 should be considered a checklist or a to-bear-in-mind caution which is appropriate for any effort to employ formal models and computerized tools in the context of policy/decision-making processes.

We should always remember that policy and decision makers are already solving their relevant decision problems. Analyzing how the decision process is currently carried out is always a prerequisite for DSS development. When the current practices are unstructured and/or non-transparent, the DSS potential increases, given that there is sufficient willingness to change. Moreover, it is known that different decision makers may lead to different conclusions for the same case. Similarly it should be expected that different results can come from the use of alternative models and DSS tools, but also from the use of the same tools by different users.

This document has benefited from the revisions of numerous colleagues in Europe and elsewhere. During the revision process many suggestions have been collected. Most of them have been implemented as far as possible in this final version, others went well beyond the limits the work and they contributed to the list of activities to be left to future – research – work that we present below.

- The various benefits of DSS, the real needs of DSS end-users and the ways in which DSS can meet these needs could be investigated in greater detail. Moreover, what benefits DSS will actually bring in terms of organization (i.e. effectiveness and efficiency improvements, learning processes), and in terms of environmental standards, or whether there is any evidence that DSS will actually produce such benefits could be further assessed. It might be useful to mention specific examples of how DSS (both within and without the context of the WFD) could help to structure complex water management problems.

- Related to the previous point is the question of knowledge demand and knowledge supply (i.e. what do policy and management organizations need in terms of know-how, know-what, know-why, etc.). This issue could be tackled by exploring the way in which knowledge demands can be delivered through DSS development and DSS application. In this regard the relevant questions are: what knowledge do river basin management organizations want or need and how do they want to access it? And what are DSS developers offering? The answer to these questions will contribute to a more accurate definition of the knowledge gaps in water policy and management.
- The report does not deal specifically with the methods needed to actually apply scientific results in decision making situations using DSS as a knowledge transfer tool, when mediators, such as private consultants, play an important role. More efforts are needed in that direction.
- Proposals for technology transfer of DSS from the research prototype to the operational tool can be further investigated. In this regard, it would be useful to tackle the subject of water policy and management organization's knowledge acquisition and management practices by ad hoc efforts.
- More should be done to analyze the potential role of academia in relation to the innovation of current management systems and the observed reluctance to use DSS. For this purpose a comparison of different systems, for example between the systems developed in-house with those developed by research projects, could be helpful. Long-term analysis of the development of tools through interactions with practitioners could also provide useful insights.
- The role of market as compared to free distribution of research products could be explored to identify more effective transfer and dissemination strategies.
- Opinions of different DSS users' groups to stress the different perspectives of users and developers could also be done by referring to the geographic distribution of DSS development and applications.

9. REFERENCES

Ajzen, I. (1985) From intentions to actions: A theory of planned behavior, in: Kuhn, J., Beckman, J. (Eds.), *Action Control: From Cognition to Behaviour*. Springer, New York, pp. 11–39.

Alter, S. (2004) A work system view of DSS in its fourth decade. *Decision Support Systems* **38**(3), 319–327.

Argent, R. M., Voinov, A., Maxwell, T., Cuddy, S. M., Rahman, J. M., Seaton, S., Vertessy, R. A. and Braddock, R. D. (2006) Comparing modelling frameworks - A workshop approach. *Environmental Modelling & Software* **21**(7), 895.

Azjen, I. (1991) The theory of planned behavior. *Organisational Behaviour and Human Decision Making Processes* **50**(2), 179–211.

Barr, S. H. and Sharda, R. (1997) Effectiveness of decision support systems: development or reliance effect? *Decision Support Systems* **21**(2), 133.

Bell, M. L., Hobbs, B. F., Elliott, E. M., Ellis, H. and Robinson, Z. (2001) An evaluation of multi-criteria methods in integrated assessment of climate policy. *Journal of Multi-Criteria Decision Analysis* **10**, 229–256.

Beven, K. (2002) Towards a coherent philosophy for modelling the environment. *Proceedings of the.Royal Society A* **458,** 1–20.

Beynon, M., Rasmequan, S. and Russ, S. (2002) A new paradigm for computer-based decision support. *Decision Support Systems* **33**(2), 127.

Blais, A.-R., Thompson, M. M. and Baranski, J. V. (2005) Individual differences in decision processing and confidence judgments in comparative judgment tasks: The role of cognitive styles. *Personality and Individual Differences* **38**(7), 1701.

Booty, W. G., Lam, D. C. L., Wong, I. W. S. and Siconolfib, P. (2001) Design and implementation of an environmental decision support system. *Environmental Modelling & Software* **16**, 453–458.

Borenstein, D. (1998) Towards a practical method to validate decision support systems. *Decision Support Systems* **23**(3), 227–239.

Bradshaw, G. A. and Borchers, J. G. (2000) Uncertainty as information: Narrowing the science-policy gap. *Conservation Ecology* **4**(1), 7.

Carlsson, C. and Turban, E. (2002) DSS: directions for the next decade. *Decision Support Systems* **33**(2), 105.

Chen, J. Q. and Lee, S. M. (2003) An exploratory cognitive DSS for strategic decision making. *Decision Support Systems* **36**(2), 147–160.

CIS (2005) *Key Issues and Research Needs under the Water Framework Directive*. Final document, comprising phase 1 and phase 2. Common Implementation Strategy for the Water Framework Directive.

Colvin, C. A. and Goh, A. (2005) Validation of the technology acceptance model for police. *Journal of Criminal Justice* **33**(1), 89.

Courtney, J. F. (2001) Decision making and knowledge management in inquiring organizations: toward a new decision-making paradigm for DSS. *Decision Support Systems* **31**(1), 17.

Davis, F. D. (1989) Perceived usefulnes, perceived easy of use, and user acceptance of information technologies. *MIS Quarterly* **13**(3), 319–340.

Davis, F. D. and Venkatesh, V. (1996) A critical assessment of potential measurement biases in the technology acceptance model: three experiments. *International Journal of Human-Computer Studies* **45**(1), 19.

Eicher, T. S. (1999) Training, adverse selection and appropriate technology: Development and growth in a small open economy. *Journal of Economic Dynamics and Control* **23**(5–6), 727.

FEEM (2005) *Report on Development and Implementation of DSS Tools in the Mediterranean Area*. Deliverable D05.5 of the NostrumDSS project.

Finlay, P. N. and Forghani, M. (1998) A classification of success factors for decision support systems. *The Journal of Strategic Information Systems* **7**(1), 53–70.

Fishbein, M. and Ajzen, I. (1975) *Belief, Attitude, Intention, and Behaviour. An Introduction to Theory and Research*. Addison-Wesley, Reading, MA.

Garrity, E. J., Glassberg, B., Kim, Y. J., Sanders, G. L. and Shin, S. K. (2005) An experimental investigation of web-based information systems success in the context of electronic commerce. *Decision Support Systems* **39**(3), 485–503.

Gelderman, M. (2002) Task difficulty, task variability and satisfaction with management support systems. *Information & Management* **39**(7), 593–604.

Gorry, G. A. and Scott-Morton, N. S. (1971) A framework for management information systems. *Sloan Management Review* **12**(3), 55–70.

Gregersen, J., Malmgren-Hansen, A., Michiel, L. and Enggrob, H. G., 2002. *Energy Environment and Sustainable Development. Requirement analysis*. Harmon-IT Report.

Harrison, E. F. and Pelletier, M. A. (2001) Revisiting strategic decision success. *Management Decisions* **39**(3), 169–179.

Hung, S.-Y. and Chang, C.-M. (2005) User acceptance of WAP services: Test of competing theories. *Computer Standards & Interfaces* **27**(4), 359.

Jakeman, A. J., Letcher, R. A. and Norton, J. P. (2006) Ten iterative steps in development and evaluation of environmental models. *Environmental Modelling & Software* **21**(5), 602.

Jeffrey, P., McIntosh, B. and Sahota, P. S. (2005) *Review Criteria and Case Study Selection Report.* Aquastress Deliverable 421 and 422.

Jiang, J. J., Muhanna, W. A. and Klein, G. (2000) User resistance and strategies for promoting acceptance across system types. *Information & Management* **37**(1), 25–36.

Jonoski, A. (2002) *Hydroinformatics as Sociotechnology: Promoting Individual Stakeholder Participation by Using Network Distributed Decision Support Systems*. PhD Thesis. TU Delft/IHE Delft, Sweets & Zeitlinger B.V., Lisse, The Netherlands.

Kahai, S. S., Solieri, S. A. and Felo, A. J. (1998) Active involvement, familiarity, framing, and the illusion of control during decision support system use. *Decision Support Systems* **23**(2), 133–148.

Kamp, J. A. L. M. (1999) Knowledge based systems: From research to practical application: pitfalls and critical success factors. *Computers and Electronics in Agriculture* **22**(2–3), 243.

Kapelan, Z., Savic, D. and Walters, G. A. (2005) *Decision Support Tools for Sustainable Urban Development.* CoDes Scoping Study:WP2 Literature Review. Centre for Water Systems, University of Exeter, UK.

Karppinen, H. (2005) Forest owners' choice of reforestation method: An application of the theory of planned behavior. *Forest Policy and Economics* **7**(3), 393.

Keen, P. G. W. and Scott Morton, M. S. (1978) *Decision Support Systems: An Organizational Perspective.* Reading, Mass., Addison-Wesley Pub. Co.

Keenan, P. B. (1998) Spatial decision support system for vehicle routing. *Decision Support Systems* **22**, 65–71.

Kukafka, R., Johnson, S. B., Linfante, A. and Allegrante, J. P. (2003) Grounding a new information technology implementation framework in behavioral science: a systematic analysis of the literature on IT use. *Journal of Biomedical Informatics* **36**(3), 218–227.

Lee, T.-T. (2004) Nurses' adoption of technology: Application of Rogers' innovation-diffusion model. *Applied Nursing Research* **17**(4), 231.

Legris, P., Ingham, J. and Collerette, P. (2003) Why do people use information technology? A critical review of the technology acceptance model. *Information & Management* **40**(3), 191.

Lilburne, L., Watt, J. and Vincent, K. (1998) A prototype DSS to evaluate irrigation management plans. *Computers and Electronics in Agriculture* **21**(3), 195–205.

Lim, J. (2003) A conceptual framework on the adoption of negotiation support systems. *Information and Software Technology* **45**(8), 469–477.

Lowry, G. (2002) Modelling user acceptance of building management systems. *Automation in Construction* **11**(6), 695–705.

Lu, H.-P., Yu, H.-J. and Lu, S. S. K. (2001) The effects of cognitive style and model type on DSS acceptance: An empirical study. *European Journal of Operational Research* **131**(3), 649–663.

Marsik, V., Horak, J., Stromsky, J. and Orlik, A. (2006) TDSS specification and design. Transcat compendium - supporting decision making in transboundary catchments.

Matthies, M., Giupponi, C. and Ostendorf, B. (2007) Environmental decision support systems: Current issues, methods and tools. *Environmental Modelling & Software* **22**(2), 123–127.

McBride, N. (1997) The rise and fall of an executive information system: A case study. *Information Systems Journal* **7**(4), 277–288.

McCown, R. L. (2002) Changing systems for supporting farmers' decisions: Problems, paradigms, and prospects. *Agricultural Systems* **74**(1), 179–220.

McFarland, D. J. and Hamilton, D. (2006) Adding contextual specificity to the technology acceptance model. *Computers in Human Behavior* **22**(3), 427.

McHaney, R. and Cronan, T. P. (2000) Toward an empirical understanding of computer simulation implementation success. *Information & Management* **37**(3), 135–151.

McKynney, D. C. (2004) *International Survey of Decision Support Systems for Integrated Water Management.*

Meppem, T. (2000) The discursive community: evolving institutional structures for planning sustainability. *Ecological Economics* **34**(1), 47–61.

Moore, R., Gijsbers, P., Fortune, D., Gregersen, J. and Blind, M. (2005) *The OpenMI Document Series. Part A Scope*. For the OpenMI, Harmon-IT Report.

Morrison, M. and Morgan, M. S. (1999) Models as mediating instruments. In: Morgan, M. S., Morrison, M. (Eds.), *Models as Mediators: Perspective on Natural and Social Sciences*. Cambridge University Press, Cambridge, UK, pp. 10–37.

Morton, A. (1990) Mathematical modelling and contrastive explanation. *Canadian Journal of Philosophy*. **16**.

Muller, A. (2003) A flower in full blossom? Ecological economics at the crossroads between normal and post-normal science. *Ecological Economics* **45**(1), 19–27.

National Research Council (1999) *Our Common Journey: A Transition toward Sustainability*. National Academy Press, Washington, D.C.

Newman, S. L. T. and Plummer, A.A. (1999) *Success and Failure of Decision Support Systems: Learning as We Go*. Conference of the American Society of Animal Science, Indianopolis, Indiana.

Nightingale, P. (2004) Technological capabilities, invisible infrastructure and the un-social construction of predictability: The overlooked fixed costs of useful research. *Research Policy* **33**(9), 1259.

Palvia, P. C. (1997) Developing a model of the global and strategic impact of information technology. *Information & Management* **32**(5), 229.

Palvia, P. C. and Palvia, S. C. (1999) An examination of the IT satisfaction of small-business users. *Information & Management* **35**(3), 127.

Palvia, S. C. and Chervany, N. L. (1995) An experimental investigation of factors influencing predicted success in DSS implementation. *Information & Management* **29**(1), 43–53.

Pielke, J. R. A. (2003) The role of models in prediction for decision. In: Canham, C., Lauenroth, W. (Eds.), *Understanding Ecosystems: The Role of Quantitative Models in Observations, Synthesis, and Prediction*. Princeton University Press, Princeton N.J., pp. 113–137.

Poon, P. and Wagner, C. (2001) Critical success factors revisited: Success and failure cases of information systems for senior executives. *Decision Support Systems* **30**(4), 393.
Power, D. J. (1997) What is a DSS? *The On-Line Executive Journal for Data-Intensive Decision Support* **1**(3),
ProGea (2004) *The WaterStrategyMan DSS - A Comprehensive Decision Support System for the Development of Sustainable Water Management Strategies*. Deliverable 21.4 of the The WaterStrategyMan Project.
Purohit, P. and Kandpal, T. C. (2005) Renewable energy technologies for irrigation water pumping in India: projected levels of dissemination, energy delivery and investment requirements using available diffusion models. *Renewable and Sustainable Energy Reviews* **9**(6), 592.
Rauscher, H. M. (1999) Ecosystem management decision support for federal forests in the United States: A review. *Forest Ecology and Management* **114**(2–3), 173–197.
Refsgaard, J. C., Henriksen, H. J., Harrar, W. G., Scholten, H. and Kassahun, A. (2005) Quality assurance in model based water management - Review of existing practice and outline of new approaches. *Environmental Modelling & Software* **20**(10), 1201.
Refsgaard, J. C., van der Sluijs, J. P., Højberg, A. L. and Vanrolleghem, P.A. (2010) Uncertainty analysis in model-based water management In: Vanrolleghem P.A. (Ed.) *Modelling Aspects of Water Framework Directive Implementation* - Volume 1. IWA Publishing, London, UK.
Ridder, D. F., Mostert, E. and Wolters, H. A. (2005) *Learning Together to Manage Together - Improving Participation for Water Management*. HarmoniCop HandBook. Institute of Environmental Systems Research, University of Osnabruck, Germany.
Rogers, E. M. (1995) *Diffusion of Innovation*. Free Press, New York.
Santhanam, R., Guimaraes, T. and George, J. F. (2000) An empirical investigation of ODSS impact on individuals and organizations. *Decision Support Systems* **30**(1), 51–72.
Sarewitz, D. and Pielke, R. A. J. (2000) Prediction in science and policy. In: Sarewitz, D., Pielke, R. A. J., Byerly, B. J. (Eds.), *Prediction: Science, Decision Making and the Future of Nature*. Island Press, Washington, DC; Covelo, California, pp. 11–22.
Scholten, H., Kassahun, A., Refsgaard, J. C., Kargas, T., Gavardinas, C. and Beulens, A. J. M. (2007) A methodology to support multidisciplinary model-based water management. *Environmental Modelling & Software* 22, 743–759.
Scholten, H. and Udink ten Cate, A. J. (1999) Quality assessment of the simulation modeling process. *Computers and Electronics in Agriculture* **22**(2–3), 199.
Schultze, U. and BolandJr, R. J. (2000) Knowledge management technology and the reproduction of knowledge work practices. *The Journal of Strategic Information Systems* **9**(2–3), 193–212.
Shaw, D. and Edwards, J. S. (2005) Building user commitment to implementing a knowledge management strategy. *Information & Management*. **42**(7), 977.
Shi, T. (2004) Ecological economics as a policy science: Rhetoric or commitment towards an improved decision-making process on sustainability. *Ecological Economics* **48**(1), 23–36.
Shim, J. P., Warkentin, M., Courtney, J. F., Power, D. J., Sharda, R. and Carlsson, C. (2002) Past, present, and future of decision support technology. *Decision Support Systems* **33**(2), 111.
Sohn, S. Y. and Ahn, B. J. (2003) Multigeneration diffusion model for economic assessment of new technology. *Technological Forecasting and Social Change* **70**(3), 251.

Stam, A., Salewicz, K. A. and Aronson, J. E. (1998) An interactive reservoir management system for Lake Kariba. *European Journal of Operational Research* **107**(1), 119–136.
Stirling, A. (2006) Analysis, participation and power: Justification and closure in participatory multi-criteria analysis. *Land Use Policy* **23**(1), 95–107.
Straub, D., Keil, M. and Brenner, W. (1997) Testing the technology acceptance model across cultures: A three country study. *Information & Management* **33**(1), 1–11.
Taylor, S. and Todd, P. A. (1995) Understanding information technology usage: A test of competing models. *Information System Research* **6**(2), 144–176.
Uran, O. and Janssen, R. (2003) Why are spatial decision support systems not used? Some experiences from the Netherlands. *Computers, Environment and Urban Systems* **27**(5), 511.
Walker, D. H., Cowell, S. G. and Johnson, A. K. L. (2001) Integrating research results into decision making about natural resource management at a catchment scale. *Agricultural Systems* **69**, 85–98.
Weiss, C. (2002) Scientific uncertainty in advising and advocacy. *Technology in Society* **24**(4), 375.
Wierenga, B. and Oude Ophuis, P. A. M. (1997) Marketing decision support systems: Adoption, use, and satisfaction. *International Journal of Research in Marketing* **14**(3), 275–290.
Wu, J.-H., Chen, Y.-C. and Lin, L.-M. (2007) Empirical evaluation of the revised end user computing acceptance model. *Computers in Human Behavior* **23**, 162–174.
Yang, H.-d. and Yoo, Y. (2004) It's all about attitude: Revisiting the technology acceptance model. *Decision Support Systems* **38**(1), 19–31.

10. APPENDICES

10.1. Appendix 1: Comprehensive list of DSS tools examined

	Tools	Developed by	Project
1	WSM DSS	National Technical University of Athens, Greece; Ruhr-University, Germany; ProGEA S.r.l., Italy	WaterStrategyMan (EU)
2	WATERWARE	Environmental Software and Services (ESS), Austria	Integrates results of the EUREKA project EU487 and related projects
3	ELBE-DSS	Research Institute for Knowledge Systems (RIKS), Maastricht NL; Dept. of Water Engineering & Management, University of Twente NL; Inst. f. Umweltsystemforschung, Univ. Osnabrück; Infram, Marknesse NL, Federal Institute of Hydrology (BfG)	Elbe-DSS
4	MDSS	Fondazione Eni Enrico Mattei (FEEM), Italy	MULINO (EU)

Appendix 1 *Continued*

5	MONERIS	Department of Shallow Lakes and Lowland Rivers; Leibniz-Institute of Freshwater Ecology and Inland Fisheries part of the Forschungsverbund Berlin e.V., Germany	
6	MERIT-DSS	The partners of the MERIT project (The guidelines for participatory use of BNs for integrated water resource management (see www.merit-eu.net). The code developer is the Machine Intelligence Group at Aalborg University, Denmark	MERIT (EU)
7	RIVERWARE	Centre for Advanced Decision Support for Water and Environmental Systems, University of Colorado, USA	
8	DELFT-TOOLS	Delft Hydraulics, The Netherlands	
9	CATCHMODS	Integrated Catchment assessment and management centre (iCAM), Canberra, Australia	
10	RAMCO	National Institute for Coastal and Marine Management and the Research Institute for Knowledge Systems (RIKS), The Netherlands	LWI-project
11	MODULUS	Research Institute for Knowledge Systems (RIKS), Maastricht, The Netherlands in collaboration with the MODULUS Consortium	MODULUS (EU)
12	MEDACTION	Department of Civil Engineering, University of Newcastle Upon Tyne, United Kingdom; Research Institute for Knowledge Systems (RIKS)	MedAction (EU)
13	ECOVISIE	Research Institute for Nature and Forest (INBO), a scientific institute of the Flemish Government in Belgium; Ministry of the Flemish Community (AMINAL), Department of Nature Brussels, Belgium; Research Institute for Knowledge Systems (RIKS)	EcoVisie

Appendix 1 *Continued*

14	DADSS	Finnish Environment Institute (SYKE), Finland	VHIMA
15	DANUBIA	Department of Earth and Environmental Sciences Chair for Geography and Geographical Remote Sensing, University of Munich, Germany	GLOWA-Danube
16	TIDDD & DTS	Gesellschaft fur angewandte Informatik mbh (FUTURE-tec Gmbh) in collaboration with Centre d'Economie et d'Ethique pour l'Environnement et le Développement (C3ED) Université de Versailles, France; the European Commission - Joint research Centre, Institute for the Protection and Security of the Citizen.	GOUVERNe (EU)
17	MIKE	Danish Hydraulic Institute (DHI), Danemark	
18	BAYES NET	Duke University, Durham, NC, USA	Neuse River, Estuary Modeling and Monitoring (ModMon)
19	GREAT-ER	Procter & Gamble ETC, Belgium; ECETOC, Brussels, Belgium	
20	AQUATOOL	Institute for Water and Environmental Engineering (IIAMA), Universidad Politecnica de Valencia, Spain	
21	HDSS (Hula Decision Support Sysrtem)	Faculty of Civil Engineering, Technion-Israel Institute of Technology, Israel	Hula Project
22	TRANSCAT-DSS	Instituto de Soldadura e Qualidade, Portugal (Project coordinator)	TRANSCAT (EU)
23	WMSS (Water Management Support System)	(Bureau de recherches géologiques et minières) BRGM, France	MEDITATE (EU)
24	WFD-EXPLORER	Delft Hydraulics, The Netherlands	
25	MOIRA DSS	ENEA Divisione Protezione dell'Uomo e degli Ecosistemi Centro Ricerche Casaccia Roma, Italy	MOIRA (EU)

Appendix 1 *Continued*

26	EIADSS (The Evaluation/Prediction of Environmental Impact Assessment for Irrigation Project)	Center for Environment and Development for the Arab Region and Europe (CEDARE), Cairo, Egypt	
27	REAL-LIFE Scale Integrated Catchment Models for Supporting Water- And Environmental Management Decisions	Water Resources Research Centre (VITUKI), Hungary	The Tisza River (EU)
28	DECIS Decision Making / Management	Consorzio Venezia Ricerche (CVR), Venice, Italy	MODELKEY (EU)
29	TWOLE	Sistemi Ambientali e Territoriali (SAT) group at Politecnico di Milano, Dipartimento di Elettronica e Informazione (DEI), Italy	TwoLe (EU)
30	DSS for the Assessment of Flood Risk	HR Wallingford Ltd, Wallingford UK (Co-ordinator)	EUROTAS (EU)
31	STREAMPLAN	International Institute for Applied Systems Analysis (IIASA)	
32	DESERT	International Institute for Applied Systems Analysis (IIASA)	
33	DSS-DROUGHT	University of Catania, Institute of Hydraulics, Hydrology and Water Resource Management (project coordinator)	DSS-DROUGHT (EU)
34	DSS for fte Management of Integrated Water Resources Systems Focused to Drought Prevention and Mitigation	University of Catania, Catania, Italy (project coordinator)	WAM-ME (EU)
35	DSS for Sustainable Ecosystem Management in Atlantic Rain Forest Rural Areas	Universidade Atlântica and Universität Hamburg (project coordinators)	ECOMAN
36	RAMFLOOD-DSS (for Risk Assessment and Management of Floods)	Centre Internacional De Metodes Numerics en l'Enginyeria (CIMNE) Barcelona, Spain (project coordinator)	RAMFLOOD (EU)
37	INFRAPLAN	Cranfield University, United Kingdom	TiGrESS

Appendix 1 *Continued*

38	SCAPT (Strategyc Catchment Analysis Planning Tool)	Cranfield University, United Kingdom	AQUADAPT
39	Decision Support System for the Effective Management of Freshwaters	ULIV - University of Liverpool, United Kingdom	EURO-Limpacs
40	DSS FOR URBAN SYSTEM ASSESSMENT	AGK - Applied Geology Karlsruhe, University of Karlsruhe, Germany	AISUWRS (EU project)
41	ADSS (Adaptive Decision Support System)	CEREVE, Centre d'Enseignement et de Recherche sur l'Eau, la Ville et l'Environnement; Common research centre UPVM, ENPC, ENGREF Ecole Nationale des Ponts et Chaussées (ENPC), France	DayWater (EU)
42	Decision Support Tools for Sustainable Water Network Management	CARE-W consortium	CARE-W (EU)
43	DSS for Rehabilitation of Sewer Networks	CARE-S consortium	CARE-S (EU)
44	DSS for Flood Prevention And Protection	Council for The Central Laboratory of The Research Councils, Department of Information Technology, Rutherford Appleton Laboratory, United Kingdom	ANFAS (EU)
45	GIBSI (Gestion Intégrée des Bassins Versants à l'aide D'un Système Informatisé)	Institut national de la recherche scientifique (INRS), Centre Eau, Terre & Environnement, Québec, Canada	
46	NELUP-DSS	University of Newcastle. Dep. Of Civil Engineering,	NERC-ESRC Land Use Programme (NELUP)
47	RAISON	NWRI Software National Water Research Institute, Canada Centre for Inland Waters, Burlington, Ontario, Canada	
48	FLOODWORKS	Wallingford Software Ltd, Oxfordshire, United Kingdom	

Appendix 1 *Continued*

49	WAMADSS (Watershed Management Decision Support System)	Centre for Agricultural, Resources and Environmental systems (CARES) University of Missouri, Columbia, USA	
50	WATERSHEDSS (Water, Soil, And Hydro-Environmental Decision Support System)	NCSU Water Quality Group at North Carolina State Universityand the Agricultural and Biological Engineering Department at The Pennsylvania State University, USA	
51	ILRDSS (Illinois River Decision Support System)	Illinois State Water Survey; Illinois Natural History Survey; Illinois State Geologic Survey, and Waste Management and Research Centre; Illinois Department of Natural Resources, USA	
52	MODSIM-DSS	Colorado State University, Civil Engineering Department, USA	
53	FLOODRELIEF-DSS	Danish Hydraulic Institute (DHI) Water & Environment, Denmark	FLOODRELIEF (EU)
54	REALM (RESOURCE ALLOCATION MODEL)	Department of Sustainability and Environment, Victoria, Australia	
55	BASINS (BETTER ASSESSMENT SCIENCE INTEGRATING POINT AND NONPOINT SOURCES)	US Environmental Protection Agency (EPA), USA	
56	IQQM (INTEGRATED QUANTITY AND QUALITY MODEL)	Department of Infrastructure, Planning and Natural Resources (DIPNR), Australia	
57	WEAP (WATER EVALUATION AND PLANNING SYSTEM)	The Stockholm Environment Institute, Hydrologic Engineering Center of the US Army Corps of Engineers funded significant enhancements.	
58	OASIS (STAKEHOLDER TOOL FOR WATER RESOURCES)	HydroLogics, Inc., USA	
59	DELFT-FEWS (FLOOD EARLY WARNING SYSTEM)	Delft Hydraulics, The Netherlands	
60	MADWICA DSS	Aquatest S.A., Geologická, Czech Republic	EUREKA E! 2721 MADWICA (EU)

10.2. Appendix 2: Presentation of short-listed DSS tools

10.2.1. CatchMODS

Primary scope/domain: The Catchment-scale Management Of Diffuse Sources (CatchMODS) model aims to assist users to identify appropriate forms of management intervention to reduce nutrient inputs to the Ben Chifley Dam Catchment (BCDC). CatchMODS also provides a vehicle for communication and interaction with managers and the catchment community.

Technical design: CatchMODS is a scenario-based modelling tool. Several of the processes represented in the SedNet model are integrated (Prosser *et al.*, 2001) and the IHACRES model (Jakeman *et al.*, 1990) is used to estimate surface and subsurface discharge in the modelling network. CatchMODS includes two nutrients submodels that simulate total nitrogen (TN) and total phosphorus (TP) export and an in-reservoir plankton-response model (specific to the Ban Chifley Dam catchment). A number of socio-economic factors are considered in the model, mostly the costs associated with the available preventative and remediation options.

The model operates with two modelling platforms. Spatial data pre-processing is undertaken using GIS-based queries for each subcatchment and river reach. The other modelling platform used is the Interactive Component Modelling System (ICMS). ICMS is used to perform the modelling and deliver an integrated Decision Support System. ICMS is an object-oriented modelling environment incorporating various models and data used in catchment management modelling. CatchMODS is based on a series of linked river reaches and associated subcatchment areas. A user interface has been developed for the BCDC application of CatchMODS and can be readily modified for new model applications.

Support provided: CatchMODS is designed to support stakeholders focus remediation on specific subcatchments and encourage sustainable management practices more broadly in the catchment. It enables development and testing of realistic land management change scenarios, and has provided valuable information on likely adoption rates for recommended management practices.

Distribution and maintenance: CatchMODS and the associated ICMS software are freely available under licence. To encourage appropriate use of the DSS its distribution is limited to those who have participated in specific CatchMODS training workshops. In the interests of transparency, the underlying code of the DSS can be accessed by all model users but modification is discouraged. Further development of CatchMODS DSS is in progress with the objective of expanding the capabilities of the modelling and increasing confidence in model outputs.

Users and applications: CatchMODS has been used to identify subcatchments that have high pollutant source and transport strengths relative to the remainder of the BCDC. Prototypes of the model have also been developed for several other catchments including the Moruya River and Colia Lake catchments, NSW and the Cox Creek catchment, SA.

Documentation: CatchMODS is distributed with a user guide including a description of the underlying model and tutorials on how to construct scenarios and examine results.

Main scientific papers:

Newham, L.T.H., A.J. Jakeman and R.A. Letcher (submitted) End-user participation in modelling for integrated catchment assessment and management: An Australian case study in participation. *International Journal of River Basin Management*;

Newham, L.T.H., A.J. Jakeman and R.A. Letcher (2006) Stakeholder participation in modelling for integrated catchment assessment and management: An Australian case study. *International Journal of River Basin Management.* **4**(3), 1–13.

Newham, L.T.H., R.A. Letcher, A.J. Jakeman, T. Kobayashi (2004) A framework for integrated hydrologic, sediment and nutrient export modelling for catchment-scale management. *Environmental Modelling & Software*, **19**, 1029–1038;

Development: The Ben Chifley Dam catchment study was assisted by the NSW Government through a grant from their Environmental Trust. The Department of Environment and Conservation (DEC) a NSW Government agency established a water quality monitoring program to support the development of the CatchMODS model. The development of CatchMODS was supported by close user participation during the case study.

10.2.2. Elbe-DSS (Pilot-Version)

Primary scope/domain: The tool's main purpose is the implementation of an IRBM tool in the German part of the Elbe river basin. The system addresses water quantity, chemical quality, highwater and navigation related aspects and ecological status of waters and the floodplain. The interaction of management objectives, external scenarios of climate, agro-economic and demographic change, and selected measures to achieve the desired state of good water quantity and quality, flooding safety and the ecological status of floodplains is investigated.

Technical design: The Elbe-DSS contains a user-oriented interface (in German), which allows handling of models and access to GIS-based information. A first, partly translated version in English is integrated. A modular, scale-related system diagram was drawn up that includes the modules "catchment", "river network", "main channel" (river plus flood-plain), and a 10-km "river section" including the floodplain) in the present version. These modules work in different spatial and temporal resolutions. The users of the DSS may select and enter planned "measures" by which they want to achieve their "management objectives". The previously defined "indicators" describe the degree of attainment of an objective. With these indicators, the user can have the consequences of his/her activity policy displayed and assessed. Additionally, the user may select previously computed "external scenarios", such as climate-change or land-use scenarios. Evaluation tools have been provided for various kinds of decision-making, e.g. risk-based for hazardous pollutant concentrations, monetary-based for engineering measures or ecological services for floodplain restoration. Different models are integrated, for example a model for the calculation of the long-term nutrient discharges from non-point sources (MONERIS) and a simulation model for wastewater pathways (point sources) and aquatic fate assessment (GREAT-ER). Further different models for hydrological dynamics were integrated. To set up all tools and models within one system the DSS-generator software GEONAMICA, developed by RIKS, and additional software was used. The Elbe-DSS is a PC application for Windows NT, 2000, XP. Under Windows NT at least 256, better 512 MB, are recommendable; under Windows XP at least 512, better 1 GB. Microsoft Excel (>2000) and Internet Explorer need to be installed at the users' PC.

The Elbe-DSS is available in a pilot-version. It will be further developed in future.

Support provided: The DSS has the purpose to facilitate the assessment of the impacts of management options. It highlights cause-effect relations and the sensitivity of different actions. To this end the DSS structures complex problems. Sectoral knowledge and available models are pooled within a basic framework. By applying a customised user interface they are made applicable in the management context. It supports the evaluation of the policy impact and scenario development. It is also intended to support participative decision making.

Distribution and maintenance: The Elbe-DSS is freely downloadable at http://elise.bafg.de/?3283. The pilot-DSS offers an operable system with an interface that is tailored to the specific requirements of users. It will be appropriate to speak of a genuine DSS for real-life decision-making when up-to-date data of the users are provided and functionalities are finally adapted. This implementation process is ongoing.

Users and applications: Potential end-users of the DSS include Internationale Kommission zum Schutz der Elbe (IKSE), Arbeitsgemeinschaft zur Reinhaltung der Elbe (ARGE-Elbe), Ministries of the "Länder" in the catchment area, Biosphere Reservations, Water management associations and authorities, Federal Waterways and Shipping Directorates (WSV), Federal Institute of Hydrology (BfG), Federal Waterways Engineering and Research Institute (BAW), Federal Ministries of Agriculture and Environment (BML, BMU), Federal Environmental Agency (UBA).

Documentation: A context-sensitive library function is generated by means of an on-line help format (WinHelp, HTMLHelp). It is structured through hypertext and internet links and allows rapid access to the desired information. For all integrated models, the scientific approaches are described, and instructions how to handle the DSS features are given. Additionally a number of documents describing different phases of the development are available on the Elbe-Informations-SystEm (ELISE) web site (http://elise.bafg.de/?3283).

Main scientific papers:

Kok, J.L. de and Kofalk S. (2006) Towards a user-oriented design of DSS for integrated river-basin management: the Elbe DSS prototype, In: Voinov, A., Jakeman, A., Rizzoli, A. (Eds). *Proceedings of the iEMSs Third Biennial Meeting: "Summit on Environmental Modelling and Software"*. International Environmental Modelling and Software Society, Burlington, USA, July 2006. CD ROM. Internet: http://www.iemss.org/iemss2006/sessions/all.html.

Matthies, M., Berlekamp, J., Lautenbach, S., Graf, N. and Reimer, S. (2006) System analysis of water quality management for the Elbe river basin. *Environmental Modelling & Software*, **21**(9), 1309-1318.

Berlekamp, J., Lautenbach, S., Graf, N., Reimer, S. and Matthies, M. (2005) Integration of MONERIS and GREAT-ER in the decision support system for the German Elbe river basin. *Environmental Modelling & Software*, In Press, Corrected Proof;

Kok, J.L. de and Huang, Y. (2005) Assessment of flood risk at various scales - The Elbe prototype DSS. In: Van Alphen, Van Beek, and Taal (Eds.), *Floods, from Defence to Management, Symposium Papers of the 3rd International Symposium on Flood Defence*, Nijmegen, The Netherlands, 25-27 May 2005, Taylor & Francis Group, London.

Development: The Federal Institute of Hydrology (BfG) initiated the design and development of the Elbe-DSS. The consortium was comprised of Bundesanstalt für Gewässerkunde, INFRAM, RIKS, Universität Osnabrück, Universität Twente. A feasibility study was completed, before the development process started. User needs were identified and refined by repeated consultation of water managers. A list of management objectives, measures, and external scenarios emerged, which was taken as the basis for the DSS development.

10.2.3. KerDST

Primary scope/domain: The KerDST (Deliberation Support Tool) is designed to function as a complement to scenario modelling and spatial (e.g. GIS based) representations and integrates the KerBabel Deliberation Matrix with its associated KerBabel Indicator Kiosk (KIK). The Deliberation Matrix permits a comparative multi-criteria multi stakeholder evaluation of scenarios for regional futures or for farm practices, and the KerBabel Indicator Kiosk helps make feasible an accessible and comprehensive documentation of information sets used in modelling and maps. The Deliberation Matrix presents, in a synthetic way, the array of judgements offered by stakeholders concerning alternative perspectives on management of the environmental resource. KerDST is applied in projects that deal with agriculture activities and choices at different spatial scales (farm scale, district or sub-catchment, and finally regional scale) and also contrasting timescales (long term planning or on-going monitoring).

Technical design: In the Deliberation Matrix, the scenarios of distinct possible futures are evaluated from distinct stakeholder perspectives. A small number of major stakeholder categories are specified, based on institutional and interview analyses. The procedure of evaluation allows for the existence of diversity within each major stakeholder category. The stakeholders' may make their evaluations of each scenario in terms of a variety of different criteria The criteria are grouped into a small number of 'baskets' corresponding to distinct governance issues. The KIK is, in itself, a generic deliberation support tool whose meta-information structure addresses the contexts of indicator use and pertinence as well as the more traditional domains of information sources. The Deliberation Matrix and the KIK are seen as naturally coupled, and the KerDST system establishes this coupling for use on-line. System' requirements: Firefox browser Macromedia Flash Player 7 and up.

Support provided: The KerBabel Deliberation Matrix provides the framework for a qualitative multi-stakeholders multi-criteria comparison of scenarios (or sites or options, etc.) by a single person or a community of users. The integrated KerDST Deliberation Matrix with associated Kiosk (KIK) allows multi-criteria evaluation to be supported by indicators and, in the "participatory option" allows judgements by each category of stakeholders to be produced as a composite outcome of multiple participants. Ideally, these tools help to develop and then maintain involvement of members of the business and wider communities around the planning or business choice questions (such as taking environmental quality into account in farming).

Distribution and maintenance: The KerDST is an on-line tool, developed with 'Open Source' conventions. Since 2006, the KerBabel Deliberation Matrix has been made available (http://kerdst.c3ed.uvsq.fr/). The integrated KerDST system of the Deliberation Matrix with associated Kiosk (KIK), is available on http://iacaprod.c3ed.uvsq.fr/kerdst2/.

Users and applications: The prototype kerDST system is in use in the SRDTOOLS EU project during 2005, and saw its prospects for the stakeholder based regional futures studies and sustainable agriculture evaluation work that AgResearch, New Zealand (www.agresearch.cri.nz/) is engaged in. The use of kerDST is now incorporated in several projects including "Choosing Regional Futures" (led by the Environment Waikato, the regional council concerned), and a set of industry sponsored projects intended to define methods for incorporating environmental and social performance considerations into assessment of agriculture strategies at farm, district and regional scales, e.g., river water quality and chemical runoff from intensive dairy farming.

Documentation: The system is distributed with a manual: "How to do it. A user's Manual for KerDST Deliberation Matrix. Non Participative KerDST using the Indicator Dialog Box." prepared by Martin O' Connor (Université de Versailles-St Quentin en yvelines - UVSQ), Victoria Reichel and Pierre Bureau (C3ED). Detailed documentation of KerDST system can be found in a series of internal C3ED documents C3ED (e.g., Bureau, Legrand, O'Connor & Reichel, 2007).

Development: During the GOUVERNe ("Guidelines for the Organisation, Use and Validation of Information Systems for Evaluating Aquifer Resources and Needs-Funded by DG RTD, 5th Framework Programme" - Contract. No. EVK1-1999-00032) and then VIRTUALIS ("Social Learning on EnVIRonmental Issues with the InTeractive Information and CommUnicAtion TechnoLogIeS" - Funded by the EC, 5th Framework Programme, 2001-2004 -Contract IST-2000-28121), the C3ED (Centre d'Economie et d'Ethique pour l'Environnement et le Développement) and partners have developed and experimented with various aspects of on-line "deliberation support tools". KerDST has been developed and tested in the context of the SRDTOOLS project ("Methods and Tools for evaluating the Impact of cohesion policies on Sustainable Regional Development", Funded by the EC 6th Framework Programme, 2005-2006, Contract No.502485).

10.2.4. MERIT DSS

Primary scope/domain: MERIT DSS is a generic integrated management tool based on the concept of Bayesian belief networks (BBNs) developed to provide an integrated and participatory water resource management methodology suitable for use throughout Europe at river basin scale. This practical and effective methodology and guidelines for construction of BBNs with stakeholder involvement enables managers to make multi-objective decisions, ensuring that stakeholders are engaged in the participatory decision-making process.

Technical design: The approach essentially consists of a decision support tool based on probability theory and Bayesian belief networks (BBNs), with guidelines for the use of BBNs as a participatory tool for Water Resource Management (Bromley, 2005). BBNs are policy development and dialogue tools able to represent the environmental system as a network of nodes linked so as to represent cause and effect within the system. Each node represents a variable and can be of any type (environmental, social, economic, etc.), and the "cause and effect" links between them can be simulated using a range of analytical techniques based on whatever data are available. The impact of a decision can be evaluated by entering the action into the relevant node (variable). This change will then have a knock on effect on all those nodes linked to it, thus the impact on the entire system can be evaluated. Results are presented as bar graphs showing the state of each variable in terms of probability distributions, thus explicitly representing the degree of uncertainty in the system. The explicit recognition of uncertainty helps decision-makers to identify more clearly the risks associated with different management strategies. A Bayesian network package called HUGIN (www.hugin.com) was used to develop the networks. The results were obtained using a procedure known as propagation. The HUGIN decision engine (HDE) performs reasoning on a knowledge base represented as a Bayesian network or an influence diagram. The HDE performs all data processing and storage maintenance associated with the reasoning process and allows application program interface or C, C++, and Java, and an Active X server. The language used in the guidelines and the documentation of the HUGIN interface is in English, but it is possible to construct the BNs using any language for the names of variables, states and documentations within the BBNs, so that the participatory use can be in any selected language.

Support provided: There is no single appropriate method of engagement replicable in all circumstances. Only the principles of engagement are replicable. Strength of BBNs are:

- BBNs are very visual; it is easy to demonstrate the way in which a system functions through the use of variables and links
- Can be used to integrate environmental, economic, social, cultural and political variables
- Permits impact of many different strategies and / or scenarios to be assessed in a short time
- Provides an excellent focus for dialogue with stakeholders
- Decisions made with a BBN are transparent when used in conjunction with stakeholder groups

Distribution and maintenance: The MERIT guidelines for the use of BBNs as a participatory tool for Water Resource Management are freely available including a free HUGIN viewer for the four case studies from MERIT. HUGIN is commercial software and maintained and used for a variety of different purposes (see www.hugin.com). The MERIT guidelines and the use of HUGIN as a participatory tool is further enhanced as a fast track existing tool in the EU research project NEWATER (2005-2008).

Users and applications: The approach was applied during the project implementation in the context of four case studies: the Havelse wellfield catchment in North Zealand (Denmark) for wich Geological Survey of Denmark and Greenland (GEUS) was the project leader and Copenhagen Energy (CE) the decision-maker and end-user in relation to the BBN case study; by the Junta Central de Regantes de La Mancha Oriental (Spain) for the Jucar catchment, by the Sperimentale Italiano Giacinto Motta S.p.a - CESI (Italy) for the Vomano catchment and by Centre for Ecology and Hydrology – CEH (UK) for the Loddon catchment. The system has been used beyond the MERIT project e.g. in Denmark for the purpose of integrating hydrological and economic models.

Documentation:

Main scientific papers:

Henriksen, H. J., Rasmussen, P., Brandt, G., Bülow, D. von and Jensen, F. V. (2007) Bayesian networks as a participatory modelling tool for groundwater protection.

Henriksen, H. J., Rasmussen, P., Brandt, G., Bülow, D. von and Jensen, F. V. (in prep.) Public participation modelling using Bayesian networks in management of groundwater contamination. *Environmental Modelling & Software.* In Press, Corrected Proof, Available online 3 April 2006.

Bromley, J. (2005). Guidelines for the use of Bayesian networks as a participatory tool for water resource management. Based on the results of the MERIT project. CEH, Wallingford, UK.

Henriksen H.J. (Ed.) (2004) *Test of Bayesian Belief Network and Stakeholder Involvement*, Ministry of Environment Geological Survey of Denmark and Greenland (GEUS) pub. (www.geus.dk)

Castelletti, A. and Soncini-Sessa R. () Topics on system analysis and integrated water resources management. Elsevier book. Chapter 3.

Development: The system developed by five MERIT's partners: Centre for Ecology and Hydrology (CEH), the Department of Computer Science, Institute for Electronic Systems, Decision Support Systems Group, Aalborg University, Denmark; the Ministry of Environment, Geological Survey of Denmark and Greenland (GEUS); the "Universidad de Castilla la Mancha" (UCLM), Spain; the "Sperimentale Italiano Giacinto Motta S.p.a", Milano, Italy (CESI). Copenhagen Energy (CE) was involved in the context of the Danish case study as the decision-maker and end-user. MERIT is a research project supported by the European Commission under the Fifth Framework Programme (Contract no.EVK1-CT-2000-00085: June 2001-May 2004).

10.2.5. MONERIS

Primary scope/domain: The MONERIS model (Modelling Nutrient Emission in River Systems) was developed in order to estimate nutrient emissions into river systems from point sources (waste water treatment plants and direct industrial discharger) and six diffuse pathways (atmospheric deposition, erosion, surface runoff, tile drainage, groundwater and paved urban areas).

Technical design: The model is conceptual, not dynamic, and designed for large river basins. It calculates total nitrogen and total phosphorus emissions from point and diffuse sources, riverine retention and resulting loads for single years or as mean for periods. The model uses information on river flow (from gauging stations), water quality (nutrient concentrations from monitoring stations), statistical data (inhabitants, sewer systems, etc.), GIS-related geographical data (stored and analysed in a GIS). It will estimate equilibrium conditions for a hydrological period. The model runs under MS-Excel 2000 or later.

Support provided: MONERIS can help managers identify proportion of pathways on the total nutrient emissions and helps to identify those to be targeted for management practices aimed at nutrient emission reduction. Combined with geographic information, it can help identify hot spots within the catchment - particular areas that, due to a combination of high potential emission and a susceptibility to efficient transport, contribute nutrients significantly more than other areas. MONERIS provides several management scenarios to reduce nutrient emissions. For example, a manager can ask by how much nutrient emissions into the river would be reduced under a scenario of erosion control.

Distribution and maintenance: MONERIS runs under EXCEL. The files are freely available by contacting Horst Behrendt at the IGB. The IGB tries to stay in close contact with people applying the model, giving explanations on the model and discussing the results.

Users and applications: MONERIS was applied in the Po River Basin (Artioli Y. *et al.* (2005), Palmeri L. *et al.* (2005)). MONERIS is used by the International Commission for the Protection of the Danube River (ICPDR) and Danube governments to assess nutrient emissions into 388 sub-basins in the Danube River Basin. It is integrated in the Elbe-DSS. MONERIS has been applied to catchments all over Europe. Within the European projects EUROHARP, BUFFER, STREAMS, DANUBS, EUROCATS and Lake Promo or by NATO-CCMS as well as national projects the model has been applied, tested and compared against different input data and other models calculating nutrient fluxes in river systems.

Documentation:

Main scientific papers:

Artioli, Y., Bendoricchio, G. and Palmeri, L. (2005) Defining and modelling the coastal zone affected by the Po river (Italy), *Ecological Modelling*. **184**(1), 55–68.

Behrendt, H. and Dannowski, R. (2005) *Nutrients and Heavy Metal in the Odra River System – Emissions from Point and Diffuse Sources, their Loads, and Scenario Calculation on Possible Changes*. Weißensee Verlag, Berlin, 353 pp.

Behrendt, H., van Gils, J., Schreiber, H. and Zessner, M. (2005) Point and diffuse emissions and loads in the transboundary Danube River Basin. – II Long-term changes. *Large rivers*. **16**(1-2), 221–247.

Palmeri, L., Bendoricchio, G. and Artioli, Y. (2005) Modelling nutrient emissions from river systems and loads to the coastal zone: Po River case study, Italy. *Ecological Modelling*. **184**(1), 37–53.

Schreiber, H., Behrendt, H., Constanttinescu, L. T., Cvitanic, I., Drumea, D., Jabucar, D., Juran, S., Pataki, B., Snishko, S. and Zessner, M. (2005) Point and diffuse emissions and loads in the transboundary Danube River Basin. – I A modelling approach.

Behrendt, H., Huber, P., Kornmilch, M., Opitz, D., Schmoll, O., Scholz, G. and Uebe, R. (2000) Nutrient emissions into river basins of Germany. *UBA-Texte*. **23**/00, 1-288.

Behrendt, H. and Bachor, A. (1998) Point and diffuse load of nutrients to the Baltic Sea by river basins of North East Germany (Mecklenburg-Vorpommern). *Water, Science and Technology*. **28**(10).

Development: The model was developed at the Leibniz Institute for Freshwater Biology and inland Fisherie Berlin in the Department of Shallow Lakes and Lowland Rivers.

(http://www.igb-berlin.de/weborg/frames_abtlg/igbabtframeset_e.htmlwas). For the development users have not been involved directly.

10.2.6. MULINO DSS

Primary scope/domain: Mulino DSS (also mDSS) is a generic decision support system developed to assist the implementation of the Water Framework Directive and the development of the River Basin Plans. It is applicable to a range of decisions - choices from predefined policy options. The main focus is on selection of measures and development of programmes of measures to meet the WFD objectives. Gaps addressed: lack of operational applicability of the IWRM concept, integration of different aspects of the problem at hand and impacts of proposed policies.

Technical design: The DSS consists mainly of methods and techniques to elicit, scrutinise and aggregate decision preferences. Various multiple criteria decision techniques (such as TOPSIS, OWA, ELECTRE, group decision techniques) are implemented and extended with techniques for sensitivity analysis and sustainability impact assessment. The DSS has interfaces to environmental models so that their predicted impacts can be incorporated in the evaluation and preference elicitation. Interface of the mDSS is built upon the DPSIR framework which guides the problem formulation and exploration. The newly released version mDSS4 is integrated within a framework (NetSyMod, see www.netsymod.eu) which contains various analyses (not encoded in the software) on the results on which the mDSS builds. The integration is facilitated in the policy assessment, taking into account various aspects of the problem at hand.

The current system is distributed in the English language only, an earlier version includes a contextual help in various languages. It requires Windows operational system, starting form Win 2000; ca. 30 MB disk space and 128MB RAM. The version 4 requires MS .NET framework to be installed. The system reads GIS data in SHAPE and ASCI formats. Implementation of OpenMI standard is envisaged.

Support provided: mDSS supports mainly the identification and evaluation of policies, and it builds on insights gained through stakeholders' involvement. The NetSyMod framework on the other hand facilitates collective learning and relationship building, and by doing so increases prospects for compromising different expectations/interests.

Distribution and maintenance: The DSS is freely available; a copy can be obtained form http://www.feem.it/mulino (under the general public license). The system is a stand-alone software working under the Windows (2000, Me, XP) operational system. The system is maintained and further developed by FEEM (www.feem.it).

Users and applications: The system was developed for water authorities dealing with water policies but the use of the software is not restricted to this user group. It was applied in a range of case studies 5 (see http://siti.feem.it/mulino/exesum/exesumm.pdf). Later the system was incorporated also in the Transcat DSS and applied in additional case studies. The software is being used for educational purposes in The Netherlands.

Documentation: The system is distributed with a tutorial (a case study developed with the help of the system), user guide and a description of the methodologies applied (decision guide).

Main scientific papers:

Giupponi, C. (2006) Decision Support Systems for implementing the European Water Framework Directive: The MULINO approach. *Environmental Modelling & Software*, In Press, Corrected Proof.

Mysiak, J., Giupponi, C. and Rosato, P. (2005) Towards the development of a decision support system for water resource management. *Environmental Modelling & Software*. **20**(2), 203–214.

Giupponi, C., Mysiak, J., Fassio, A. and Cogan, V. (2004) MULINO-DSS: a computer tool for sustainable use of water resources at the catchment scale. *Mathematics and Computers in Simulation*. **64**(1), 13–24.

Development: The system was developed by an international research consortium (9 institutions from 5 European countries, see http://siti.feem.it/mulino/home/consort.pdf). The development was funded by the EC under the FP5 (Energy, Environment and Sustainability Development programme, key action water). The development was driven by a close involvement of 13 policy makers/stakeholders form 7 various water authorities who helped to specify the requirements and, later, applied the system.

10.2.7 RiverWare

Primary scope/domain: RiverWare provides a modelling environment to meet the modelling needs of managers and operators of river and reservoir systems. It allows the management of multiple system objectives as water supply, flood control, navigation recreation, fish and wildlife habitat and hydropower production. RiverWare simulates the hydrologic processes of the river and reservoir system. In simulation the solution is driven by logical operating rules.

Technical design: RiverWare is a general river basin modelling tool framework with a user-friendly point-and-click GUI. The basic building blocks of a RiverWare river basin model are objects which represent the features of the river basin such as reservoirs, river reaches, consumptive users, canals, shallow groundwater aquifers, etc. The objects (icons on the workspace) contain their own data and physical process algorithms. The selection of solution algorithms are all provided through a user-friendly interface (currently in English only). In simulation, the solution is driven by input data and/or prioritised logical operating rules. In optimisation, prioritised objectives/constraints are solved with a linear goal programming solution using CPLEX (must be licensed separately). RiverWare automatically generates the physical constraints and linearisations. Operating rules and objectives are formulated in the RiverWare Policy Lanaguage (RPL), an interpreted language, in a syntax-directed editor. RiverWare also models water ownership and accounting and can allocate flows to satisfy prioritised water rights. RiverWare has numerous features for ease of use such as multiple run management for generation of probabilistic results, linking with any external data source, spreadsheet-like views of data, and sophisticated analysis facilities.

RiverWare is a c++ application which runs on Sun Solaris (Unix) workstation with Solaris 2.7 or higher operating system, or Windows NT / 2000 / XP.

Support provided: The RiverWare software can be applied to river basins for operational scheduling, forecasting and planning applications, for evaluation of new projects, yield studies, and policy analysis and evaluation. Participating agencies and stakeholders can share models, modelling scenarios and model results, facilitating prediction of proposed (or alternative) policy outcomes and communication among agencies and stakeholders. RiverWare is used in interactive meetings with stakeholders to explore possible outcomes of proposed alternatives. Uncertainty analysis is achieved by probabilistic representation of the outputs of multiple (stochastic) hydrologic sequences enabled by RiverWare's Multiple Run Manager and the Graphical Policy Analysis Tool (GPAT). Policy options are identified by use of the interactive policy language that controls simulation, or via the Goal Programming module which considers hard and soft policy constraints and objectives to determine best operations.

Distribution and maintenance: By agreement with the sponsoring agencies, RiverWare is licensed through the University of Colorado Technology Transfer Office. License revenues contribute to software maintenance. RiverWare is currently licensed to more than 60 water management agencies, consultants, universities, research institutes and foreign entities. User and installation support are provided and a web-based bug-tracking facility allows users to log problems directly CADSWES offers "Introduction to Simulation Modelling" and "Rulebased Simulation Modeling", both 3-day training classes. An annual Users meeting is organised. One or two major releases are made each year. RiverWare is under constant enhancement and new development to meet the evolving needs of users.

Users and applications: The tool has been applied to numerous basins by the sponsoring agencies (the U.S. Bureau of Reclamation (BOR), the Tennessee Valley Authority (TVA) and the U.S. Army Corps of Engineers) and by the many water management agencies and consultants who license the software. RiverWare is used on a daily and annual basis for operations and planning on the Colorado River, the Rio Grande, the Tennessee Valley and many other river systems. New applications are frequently initiated.

Documentation: RiverWare is provided with an online Help. Additional documentation is provided with the training manuals.

Main scientific papers:

Zagona, E., T. Magee, D. Frevert, T. Fulp, M. Goranflo and J. Cotter (2005) RiverWare. V. Singh & D. Frevert (Eds.), *Watershed Models*, Taylor & Francis/ CRC Press: Boca Raton, FL, 680 pp.

Magee, T. M., H. M. Goranflo and S.H. Biddle (2002) Optimizing daily reservoir scheduling at TVA with RiverWare. In: *Proceedings of the Second Federal Interagency Hydrologic Modeling Conference.* Las Vegas, NV.

Eschenbach, E. A., T. Magee, E. Zagona, M. Goranflo and R. Shane (2001) Goal programming Decision Support System for multiobjective operation of reservoir systems. *Journal of Water Resources Planning and Management*, ASCE **127**(2), 71–141.

Magee, T. M., E. A. Zagona and D. Frevert (2001) Operational policy expression and analysis in the RiverWare modeling tool. In: *Proceedings of the ASCE World Water & Environmental Resources Congress*, Orlando, FL.

Zagona, E. A., T. J. Fulp, R. Shane, T. Magee and H. M. Goranflo (2001) RiverWare: A generalized tool for complex reservoir systems modeling. *Journal of the American Water Resources Association*, AWRA **37**(4), 913–929.

Zagona, E. A., T. J. Fulp, H. M. Goranflo and R. M. Shane (1998) RiverWare: A general river and reservoir modeling environment. In: *Proceedings of the First Federal Interagency Hydrologic Modeling Conference*, Las Vegas, NV.

Development: RiverWare is under continued development at CADSWES, funded by the R&D sponsors (TVA, Reclamation and Corps of Engineers). It is the intention of CADSWES and the sponsors to maintain the software in the future and provide enhancements as requested. The RiverWare Viewer is a cost-free version that permits viewing and post-processing of model results.

10.2.8 TwoLe

Primary scope/Domain: TwoLe supports the realisation of River Basin Management Plans as foreseen by the Water Frame Directive (2000/60/EC), in accordance with the IWRM (Integrated Water Resources Management) paradigm, the framing and exploration of Multi-Objective problems, with multiple Decision Makers and with uncertainty; the integration of long-term planning and short-mid-term management within the decision making process

Technical design: TwoLe supports multi-objective decision-making processes (MODSS, Multi Objective Decision Support System) and is structured on two integrated levels, which emulate the structure of a Water Agency: the planning and the management levels. It follows the phases of the *Participatory and Integrated Planning (PIP) procedure* (see References below) and provides methods and tools to support the activities foreseen within each phase. It is based on a multi-channel architecture which takes the form of a series of modules interacting in a single environment and makes available various services through different media to respond to the user's requirements.

The planning level is constituted of a stand-alone application which offers the users the support tools for the various phases of planning. In particular, the conceptualisation phase allows the user to integrate different models into the system: those present within the software environment can be supplemented by others possibly remotely available, and these can interact with the system through ad hoc XML data exchange protocols. The GUI, implemented in JAVA, controls dedicated computation engines (of identification of PARMAX class models, of optimization of policies through Dynamic Programming, and of deterministic, Markovian or Monte Carlo simulation) implemented in C++ and allows the use of a library of functions, expandable by the user, for the calculation of evaluation indicators. This library is written in *MASTRAVE*, a restriction of the MATLAB® syntax specifically created to be compatible with its open-source clone OCTAVE. To explicitly represent the possible conflicts between the indicators and to make clear and transparent the subjective elements introduced by stakeholders, TwoLe provides the *Multi Attribute Value Theory* (MAVT), the *Analytic Hierarchy*

Process (AHP) and the ELECTRE family of methods. The negotiation process is then supported by two approaches: the step-by-step approach, in particular negotiations on weights, and negotiations on ranking, based on the Pareto Race, a technique which drives the stakeholders in search of possible compromises in the exploration of efficient solutions.

The management level receives the reference framework (i.e., decisions, data and models) from the planning level (a priori hypothesis) via the sharing of the data structure. This part of the application thus appears as a web-service. Lastly, the results of the experiments can be compared in appropriate synoptic reports and made available. The experimentation environment is strictly linked to that of remote negotiations, which take the form of video chat. The components of TwoLe exposed to external access are protected by implementing appropriate ACL (Access Control Lists) which are extremely granular: they in fact allow the attribution of rights of writing and/or reading at the level of individual service and individual user. The creation of a TwoLe interface to GIS (in particular GRASS) is still in the design phase.

Support provided: TwoLe assists the planning procedure in all its phases. In fact it permits the user to: construct evaluation hierarchies, identify stochastic inflow models, and construct networks of interconnected models to represent water systems, design of alternatives, normative and structural actions, evaluate performance for different scenarios, compare them from different points of view, and assist negotiations for the search for compromise alternatives. TwoLe implements the planning policy in the short-medium term management, assisting the DM in the choice of the decision to implement and allowing remote negotiations among different users.

Distribution and maintenance: The software does not have a public release yet. It is being engineered and it is expected to be available in a beta version in spring 2007. Its development can be monitored on www.twole.info.

- HW requirements: Dedicated server (ca. 1GB ram), unlimited clients
- OS requirements: OpenSource Linux operating system (kernel >=2.4), Unix (SUN Solaris)
- SW requirements: Apache >= 2.2, PHP >= 5, Postgresql >= 7.3, Mediawiki >= 1.7

For possible applications, you are asked to contact the development team (mailto:feedback@twole.info)

Users and applications: A preliminary version of the planning module was used in the international project Verbano Interreg II-UE. It was successively applied to Lake Iseo and to the Vomano water system. A new version of TwoLe is currently being developed and tested on Lake Como. The management module is being developed and, at the same time, applied once again to Verbano. TwoLe has supported the planning negotiation process between government bodies, Regulation Consortia, Draining and Irrigation Consortia, hydropower companies, associations of various kinds and citizens.

Documentation: The documentation utilizes different supports, ranging from canonical pdf manuals to hypertexts, and reflects the modular structure of the support system. The easiest level is devoted to stakeholders and DMs and is based on examples and case studies periodically updated. The more complex level, tailored for experts and analysts, allows them to use the support system and to upgrade it with new models, indicators, etc. The third level (for advanced analysts and programmers) is the very code of TwoLe which has been formulated in such a way as to allow its internal structure to know information about itself and to provide it to the user. The web site of TwoLe constitutes the natural interface among these levels and permits the systematic upgrading of the documentation.

Main scientific papers:

Soncini-Sessa, R., A. Castelletti and E. Weber (2007) *Integrated Water Resource Management. Theory.* Elsevier, Amsterdam, NL.

Soncini-Sessa, R., F. Cellina, F. Pianosi and E. Weber (2007) *Integrated Water Resource Management. Practice*. Elsevier, Amsterdam, NL.

Castelletti, A. and R. Soncini-Sessa (2006) A procedural approach to strengthening integration and participation in water resource planning. *Environmental Modelling & Software*, **21**(10), 1455–1470.

Soncini-Sessa, R., A. Castelletti and E. Weber (2003) A DSS for planning and managing water reservoir systems, *Environmental Modelling & Software*, **18**(5), 395–404.

Soncini-Sessa, R., A.E. Rizzoli, L. Villa and E. Weber (1999) TwoLe: A software tool for planning and management of water reservoir networks. *Hydrological Sciences Journal*, **44**(4), 619–631.

Soncini-Sessa, R., A. Nardini, C. Gandolfi and A. Kraszewski (1990) Computer aided water reservoir management: a prototype two level DSS. In: *Proceedings of the NATO ARW on Computer Aided Support Systems in Water Resources Research and Management.* September 23-28 1990. Ericeira, Portugal.

Development: The MODSS TwoLe prototype was realised in the course of the Verbano Interreg II-UE project. The improvement of the first version of the MODSS continued in the Sebino project, which was financed by Fondazione Cariplo and within the MERIT (Fp5-UE) project. TwoLe is the System supporting the PIP procedure adopted within the WP3 of the Harmoni-CA project. The second version of the MODSS is being realized within the project Twole-A/B, a system for the Planning and Management of Water Resources; Pilot Application to the Ticino and Adda Basins, financially supported by Fondazione Cariplo.

10.2.9. WaterWare

Primary scope/domain: WaterWare is a modular, interactive water resources management information system, designed to support the implementation of the WFD and similar national legislation. Its purpose is to assist government agencies, river-basin commissions, etc., in decision-making for the efficient management of water resources in terms of both quantity and quality.

Technical design: The system combines geographical information systems, database technology, modelling techniques, optimisation procedures and expert systems. It includes a multimedia user interface with Internet access, a hybrid GIS with hierarchical map layers, object data bases, time series analysis, reporting functions, and embedded expert systems for estimation, classification and impact assessment tasks, and a hypermedia help- and explain system. Real-time data management, simulation and optimisation modelling, with data assimilation, forecasting, and reporting, and support for operational management can be provided with a real-time rule-based expert system.

WaterWare is available for servers running any kind of OpenSource Linux operating system, as well as various UNIX implementation (e.g., SUN Solaris). Other UNIX platforms can be supported on request. Any networked client (standard PC) supporting a standard web browser (MS Internet Explorer, Mozilla, Firefox) with XVGA resolution or better is supported. Auxiliary tools manage user requirements and stakeholder preference structures.

Support provided: The system supports monitoring, water allocation, pollution control and environmental impact assessment tasks.

Distribution and maintenance: A single-server, unlimited-clients version (open-ended, non-transferable binary license), include an installation, training, and first-year support package.

Users and applications: The DSS was developed through a series of applications (River Thames in England, Lerma-Chapala basin in Mexico, West-Bank and Gaza in Palestine, Kelantan River in Malaysia and the Yangtze River in China). Applications around the Mediterranean in the EU-sponsored projects SMART and OPTIMA include river basins in Cyprus, Turkey, Lebanon, Jordan, Palestine, Egypt, Tunisia and Morocco.

Documentation: An On-line Reference Manual is available at the web address http://www.ess.co.at/MANUALS/WATERWARE/R51reference.html.

Main scientific papers:

Fedra, K. and Jamieson, D. G. (1996) The "WaterWare" decision-support system for river-basin planning. 2. Planning capability. *Journal of Hydrology*. **177**(3–4), 177–198.

Jamieson D. G. and K. Fedra (1996) The "WaterWare" decision-support system for river-basin planning. 3. Example applications. *Journal of Hydrology*. **177** (3-4), 199–211.

Jamieson, D. G. and Fedra, K. (1996) The "WaterWare" decision-support system for river-basin planning. 1. Conceptual design. *Journal of Hydrology*. **177**(3-4), 163–175.

Development: The DSS is being developed on the basis of concepts and results of the EUREKA project EU487. The initial case study in this project addressed the Thames basin in England, and was sponsored by Thames Water International Ltd.

10.2.10. WaterStrategyMan DSS

Primary scope/domain: The primary aim of the WaterStrategyMan DSS (WSM DSS) was the formulation and evaluation of alternative (improved) water management strategies for water deficient regions in the context of implementing the European Water Framework Directive 2000/60/EC. The system simulates and analyses the behaviour of water resource systems under different scenarios of water availability and demand, as well as alternative allocation schemes. The domains addressed by the WSM DSS include: (a) hydrology modelling, (b) water allocation under scarcity conditions, (c) simplified water quality modelling, (d) assessment of financial, environmental and resource costs, and (e) evaluation of alternative allocation/management schemes, on the basis of indicators and multi-criteria analysis.

Technical design: The WSM DSS was developed as a stand-alone GIS-based package with the objective to emphasise the conceptual links between the different components and aspects of water resource system management, instead of merely focusing on a detailed description of the physical systems and related phenomena. Its main functions have been framed under the Driving Forces-Pressures-State-Impacts-Responses (DPSIR) structure of indicators. The core components of the tool are: (a) the database; (b) the object model linked to mathematical models for water allocation, quality and economic estimations, (c) a logical coordination unit, responsible for the communication with external models, and (d) the user interface that allows for the definition of data and simulation parameters, and the presentation of results through customisable charts, tables and maps. The operational framework of the DSS is based on the concept of a water management scheme (WMS), defined as a set of scenarios for variables that cannot be directly influenced by the decision-maker and the application of one or more water management interventions. User interaction with the DSS falls under three functional groups, accessed via a hierarchical navigation tree: (1) base case editing, allowing for the editing and introduction of new data for the reference year; (2) creation of WMSs, providing the capabilities for defining scenarios on water availability and demand, definition of strategies and visualisation of results, and for conducting a parametric economic analysis; and (3) evaluation of alternative schemes, based on a multi-criteria analysis, which takes into account the entire simulation horizon, based on the computation of statistical criteria for reliability, resilience and vulnerability for regionally applicable indicators.

The DSS has been developed in Visual Basic .NET using the Arc Objects COM technology by ESRI. The current version is distributed in the English language only. The system requires Windows operating system (2000, XP or higher), approx. 50 MB of free disk space and 256 MB of RAM. Prerequisites for the installation are the MS .NET framework 1.1 and Arc Info v. 8.x.

Support provided: 1) Assessment of the state of a water resource system in terms of sources, usage, water cycles (pathways) and environmental quality. 2) Evaluation of the impacts of different water management options, on the basis of alternative scenarios and policies.

Distribution and maintenance: A demo of the WSM DSS is available at: http://www.geomin.unibo.it/hydro/WSM/DSS_demo.htm. The DSS can be made freely available upon contact with the National Technical University of Athens, and is subject to a user license agreement. The system is being further maintained and developed by the Environmental and Energy Management Research Unit of the School of Chemical Engineering of the NTUA (http://environ.chemeng.ntua.gr/).

Users and applications: The system has been developed for addressing the needs of decision makers in arid or semi-arid regions and river basins, characterised by current or future water stress problems. Seven case studies have been developed thus far, for the arid or semi-arid regions of the island of Paros, Greece, the region of Ribeiras do Algarve, Portugal, the region of Limassol, Cyprus, the island of Tenerife, the Canary Islands, Spain, the areas of Tel Aviv and Arava, Israel, and the area of the Belice River Basin, Sicily, Italy.

Documentation: A manual can be downloaded from the WaterStrategyMan web site, at http://environ.chemeng.ntua.gr/wsm/Default.aspx?t=60.

WaterStrategyMan Project (2004). The WaterStrategyMan DSS. A Comprehensive Decision Support System for the Development of Sustainable Water Management Strategies. Bologna, Publishable report of the project "Developing strategies for regulating and managing Water Resources and Demand in Water Deficient Regions", EU DG Research, Contract n° EVK1-CT-2001-00098.

Main scientific papers:

Todini E., Shumann A. and Assimacopoulos D. (2006) The WaterStrategyMan Decision Support System. In: *Water Management in Arid and Semi-arid Regions: Interdisciplinary Perspectives*. Edward-Elgar Publishing Ltd.

Assimacopoulos D., Barraqué B., Berland J.M., Feinerman E., Katsiardi P. and Manoli E. (2005) Estimation of the level of cost recovery of different scenarios of water allocation in arid areas / a proposition for an easy-to-implement approach. In: Proceedings *Second International Workshop on Implementing economic analysis in the Water Framework Directive*. Paris, France. February 2005.

Manoli E., Katsiardi P., Arampatzis G. and Assimacopoulos D. (2005) Comprehensive water management scenarios for strategic planning. In: *Proceedings 9th International Conference on Environmental Science and Technology*. Rhodes, Greece. September 2005. pp. 913–920.

Assimacopoulos, D. (2004) An Integrated Decision Support System for the evaluation of water management strategies. In: *Proceedings IDS-Water Europe Web Conference*, May 2004.

Manoli, E., Arampatzis, G., Pissias, E, Xenos, D. and Assimacopoulos, D. (2001) Water demand and supply analysis using a spatial decision support system. *Global NEST: The International Journal*, **3**(3), 199–209.

Development: The WaterStrategyMan DSS was one of the main outputs of the WaterStrategyMan Project, "Developing Strategies fir Regulating and Managing Water Resources and Demand in Water Deficient Regions", financed by the EC through the Fifth Framework Programme, in the framework of the Key Action Sustainable Management and Quality of Water within the Energy, Environment and Sustainable Development Specific Programme (Contract No: EVK1-CT-2001-00098). All WSM partners were actively involved in the DSS formulation, while the technical development was undertaken by the National Technical University of Athens (Greece), the Ruhr-University Bochum (Germany) and ProGEA S.r.l. (Italy).

10.2.11. WFD-Explorer

Primary scope/domain: The WFD-Explorer performs a number of functions related to the implementation process of the WFD. First of all it provides a platform to integrate and disclose ecological information on water bodies, including a common knowledge base on the effectiveness of measures. It offers to water managers and stakeholders the opportunity to learn about the relationships between the objectives pursued, the measures that might be taken and the impacts of such measures (including costs). It also contributes to the harmonisation of the way in which ecological objectives are determined in the various river basins. As such it promotes the coordination of the national and regional WFD-processes. The WFD-Explorer aims to bridge the gap between the information needs of the (interactive) policy process and the results from detailed studies on water systems and processes.

Technical design: The base version of the WFD-Explorer has a modular set-up and consists of a multilingual user-interface and a computational core which are coupled through code as well as an MS Access Database. The computational core includes a balance of water and substances for the water bodies in the river basin. Most data of the WFD-Explorer are inserted in databases. Knowledge rules for ecological quality describe how the 'scores' for the four different biological quality elements (phytoplankton, macro phytes, macro fauna and fishes) depend on the values of the steering variables (hydromorphological and water quality characteristics). Knowledge rules have also been derived for a number of different types of measures. These describe how characteristics of the water system or steering variables will change because of the execution of particular measures.

The current system is stand-alone software and requires Windows XP or NT operational system; ca. 100 Mb hard disk and at least 512 Mb RAM.

Support provided: the WFD-Explorer supports water managers with the development of river basin management plans, in particular with respect to the determination of cost-effective strategies for the realisation of both ecological and chemical objectives. The WFD-Explorer aims to contribute to the discussion and communication of strategies being developed.

Distribution and maintenance: In September 2006 the base version of the WFD-Explorer has been made available to water management authorities and consultancy firms in The Netherlands. The WFD-Explorer is freely available and can be downloaded from the website www.krw-verkenner.nl. The knowledge base with the knowledge rules is protected to promote consistency among the applications for the different water bodies. If new knowledge rules are developed in particular projects, then these new knowledge rules (after peer review) are incorporated in the knowledge base and become available to all users. A maintenance organisation is being set-up for the coming years. This organisation is financed by both the Ministry of Public Works and Water Management and the Association of Water Boards.

Users and applications: A number of water boards and river basin districts are currently in the process of developing applications of the WFD-Explorer for their area. Other water boards are planning to do the same. Also applications outside the Netherlands are in progress for Belgium and Romania.

Documentation: The system is distributed with a tutorial, a user manual and a description of the methodology adopted. These include fact sheets for the different types of measures as well as fact sheets for the ecological knowledge rules of different water types. The documentation is currently in the Dutch language. However an English version of the user manual will become available in the coming months.

Main scientific papers:

Most, H., van der, Jeuken, A. and Schijndel, S., van (2006) WFD-Explorer: a planning kit to support the implementation of the water framework directive. Paper presented at *7th International Conference on Hydroinformatics HIC2006*. Nice, France.

Development: The WFD-Explorer has been developed in an incremental way. The prototype has been further developed into a base version which can be easily applied elsewhere in The Netherlands and abroad. This further development was carried out in conjunction with a number of ongoing planning studies in The Netherlands.

The WFD-Explorer has been developed by a consortium with representatives of different Water Boards, the STOWA (Foundation for Applied Water Research), the RIZA (Institute for Inland Water Management and Waste Water Treatment) and a number of knowledge institutes and consultancy firms. The project is part of the government-sponsored research program 'Living with water'.

10.3. Appendix 3. Questionnaire for DSS end users

A) DEVELOPMENT, ACQUISITION AND APPLICATION CONTEXT

A1) Is [DSS] the first decision support system you have used?

It is the first DSS I used ☐ I used another DSS before ☐

What was the system(s) you used before? ____________

A2) What was the main motivation/reason for development/acquisition of [DSS]?

Curiosity, interest in exploring how to rationalise decisions.............................. ☐
Proposed partnership from research institution
(e.g. involvement in an exploratory research project).. ☐
Emerging needs, inability of the previous system to support
increasingly complex decisions.. ☐
Technological innovation (e.g. updating the system
to meet new software/hardware standards)... ☐
Formal enforcement in your institution beyond your influence.......................... ☐

Other, please specify ____

____

____

____

A3) What outcomes did you aim at as you applied the [DSS]?

Better understanding - explaining the causes of the problem at hand.................☐
Satisfying specific regulatory obligations and requirements.............................. ☐

Facilitating stakeholders' involvement.. ☐
Eliciting perspectives form various actors about the problem at hand............. ☐
Identifying policy/project alternative options.. ☐
Evaluating previously adopted policy/design options..................................... ☐
Making a previous policy decision transparent, or demonstrating
that the decision was correct.. ☐
Innovating, including testing of new governance styles.................................. ☐
Operational management.. ☐
Other, please specify: ______________________________

A4) The implementation of the DSS imposed changes to the process of decision-making in your organisation	strongly agree	agree	neither agree nor disagree	disagree	strongly disagree
	☐	☐	☐	☐	☐

A5) Does [DSS] integrate other software instruments (such as GIS, database management system, models) previously used in your institution? If yes, please specify: ______________________________

____

____

The integration with other systems was important for the decision to develop/acquire the system	strongly agree	agree	neither agree nor disagree	disagree	strongly disagree
	☐	☐	☐	☐	☐

B) YOUR PERSONAL BACKGROUND

B1) What is your area of expertise and the role in your institution? Please specify:
Area of expertise: ______________________________
Your role in the institution: ______________________________

B2) Your involvement in the development/acquisition included:
- Specification of the requirements, identification of the current and potential needs.. ☐
- Adaptation of the system to fit the data flow and systems/databases in your institution.. ☐
- Testing and evaluation of the prototypes at various development stages.......... ☐
- Application of the system in practical situations before the system/acquisition was completed... ☐

- Consultation with other potential users, or parties involved or affected by future application of the system.. ☐
- Training of technicians.. ☐
- None of the above.. ☐

Other, please specify____

B3) Since when have you been using the [DSS] system and how frequently? ____
____
____

B4) Are you a direct user of the system (i.e. you interact with the system installed on your PC)? Yes ☐ Not ☐

	strongly agree	agree	neither agree nor disagree	disagree	strongly disagree
B5) It was difficult to get familiar with the system	☐	☐	☐	☐	☐
B6) Your role in the process of decision making changed because of the adoption of [DSS]	☐	☐	☐	☐	☐

If yes, how? ____
____
____

C) IMPLEMENTATION, USE AND MAINTENANCE

C1) What was the importance of the [DSS]' results for the policy decision to be taken?

[DSS]' results informed the decision (choices of the policy) but the choice has been taken independently of the results............................... ☐

[DSS]' results have been (more or less) directly taken up by the decision/policy choice... ☐

[DSS] was useful in a broader context, facilitating the planning process and learning among the actors, contributing to finding a compromise among different expectations and interests................................ ☐

Other, please specify____

C2) Is the system described above still maintained and further developed within your institution or in collaboration with the developer? .... Yes ☐ not ☐

C3) Have you applied the [DSS] in more than one decision context?... ☐ ☐

If yes, how many times? ____

____

	strongly agree	agree	neither agree nor disagree	disagree	strongly disagree
The re-use of the system was simple	☐	☐	☐	☐	☐

Please, insert your comments ____

____

____

D) BENEFITS AND SATISFACTION

	strongly agree	agree	neither agree nor disagree	disagree	strongly disagree
D1) The [DSS]' implementation in policy-making improved the usual/traditional approach of your Institution/Firm.	☐	☐	☐	☐	☐
D2) You are satisfied with the various components of the system (Leave blank if not applicable)...................................	☐	☐	☐	☐	☐
Database management system..............	☐	☐	☐	☐	☐
Simulation model(s) (please specify below if different models are used)...........	☐	☐	☐	☐	☐
User interface.......................................	☐	☐	☐	☐	☐
Module facilitating the evaluation/ranking of policies......................................	☐	☐	☐	☐	☐
Module responsible for documentation and presentation of the results..............	☐	☐	☐	☐	☐
Interactive exploration of the model's suitability...	☐	☐	☐	☐	☐
Role attributed to various interests groups (stakeholders)............................	☐	☐	☐	☐	☐

What were the most effective components and for what reason/s? ____________

____

and what were the main limitations and for what reason/s? ________________

____

Additional comments ____

____

D3) The use of the system or the quality of its components were contested by those who were affected by or involved in the decisions made with the help of the DSS..	strongly agree	agree	neither agree nor disagree	disagree	strongly disagree
	□	□	□	□	□

If yes, what were the main contestations? ______________________________

____

____

D4) Do you have any recommendation for the improvement of the development of similar systems in the future?____

10.4 Appendix 4. Results of the Questionnaire for DSS End Users

A) DEVELOPMENT, ACQUISITION AND APPLICATION CONTEXT

A1) Is [DSS] the first decision support system you have been using?
It is the first DSS I used: 19 I used another DSS before: 5
What was the system(s) you used before?
R1: SWAT, MFA;
R6: No WFD system;
R14: Spreadsheets/Experience;
R15: Site-specific; Fortran-Based;
R16: HEC-5;
R20: Mike SHE /mudflow;
R26: Environment Waikato and Ag research make extensive use of modelling and decision support, e.g., choice of farming regimes. However, there is less experience with DST for territorial futures as in the current "Choosing Regional Futures" project.

A2) What was the main motivation/reason for development/acquisition of [DSS]?
Curiosity, interest in exploring how to rationalise decisions...................... 5
Proposed partnership from research institution (e.g. involvement in an exploratory research project.. 14

Emerging needs, inability of the previous system to support increasingly complex decisions..........	9
Technological innovation (e.g. updating the system to meet new software/ hardware standards..........	2
Formal enforcement in your institution beyond your influence..........	2

Other, please specify:

R3: Implementation of the Water Framework Directive: Emission register for diffuse sources;
R4: Having a tool to predict the effects in change of water quality and its effects on ecology and because of the knowledge database of dosis-effect relations (there is nothing simular in the Netherlands).

A3) What outcomes did you aim for as you applied the [DSS]?

Better understanding - explaining the causes of the problem at hand	16
Satisfying specific regulatory obligations and requirements	11
Facilitation of stakeholders' involvement	18
Eliciting of perspectives form various actors about the problem at hand	5
Identification of policy/project alternative options	17
Evaluation of previously adopted policy/design options	7
Making transparent a policy decision previously taken, or demonstrating that the decision was correct	11
Innovation, including testing of new governance styles	7
Operational management	9

Other, please specify

R2: MONERIS is used in our institute (Department of Water and River Basin Management, University Karlsruhe) in the framework of a research project with the aim of modelling heavy metal emissions into river systems. This project is funded by the Umweltbundesamt (Federal Environment Agency of Germany);
R3: Status quo-investigation on share of different diffuse nutrient input pathways into surface waters

	strongly agree	agree	neither agree nor disagree	disagree	strongly disagree
A4) The implementation of the DSS imposed changes to the process of decision making in your organisation	3	6	9	2	3

A5) Does [DSS] integrate other software instruments (such as GIS, database management system, models) previously used in your institution? If yes, please specify:
R1: no
R2: GIS (ESRI ArcView, ArcGIS), database management (UDIS)
R3: GIS: data on nutrient balances (administrative units), river catchments, DEM, Corine Landcover, River network, data on water quality, flow etc
R4: Yes, the WFD explorer will be expended with a communicationtool that make it possible to make integrate choices in combination with water quantity goals and ecologigal goals for nature and in the future we want to integrate the WFD-Explorer with a model for surfacewaterquantity and –quality (Sobek)
R5: GIS information (waterbodies), databases on environmental information (STONE, ERC), hydrological model information (MOZART, NAGROM)
R8: YES; GIS: river cadaster/model: VIDRA/SWAT
R9: GIS For Caia Water Board District
R10: Gis and ground flow model
R13: The WSM-DSS tool uses GIS capabilities, to handle both vector and raster information and also includes the ProWam module, devote to the generation of water availability scenarios
R14: RiverWare integrates with our hydrologic database (HDB) (an Oracle database), and Modflow integration is underway – GIS integration is planned.
R15: yes we use an Oracle based DB and other models including HEC-RAS; we also have a real time data colllection network, all integrated through the DB with RW models
R16: HEC-DSS (data-storage system)
R18: GIS, Hydrological Models
R19: Yes, obviously in order to get the tool we had to use a great quantity of information from a lot of sources
R21*: No
R22*: Database
R23*: Will acquire data from our database
R24*: yes GIS
R25: In this case study, CatchMODS integrates specifically a reservoir-plankton-model primarily developed by our institution to understand the likely effects of catchment management on the reservoir phytoplankton blooms.
R26: The kerDST system is designed to function as a complement to scenario modelling and spatial (e.g., GIS based) representations. For example the Deliberation Matrix permits a comparative multi-criteria multi stakeholder evaluation of scenarios for regional futures or for farm practices, and the KerBabel Indicator Kiosk helps make feasible an accessible and comprehensive documentation of information sets used in modelling and maps.

The integration with other systems was important for the decision to develop/ acquire the system	strongly agree	agree	neither agree nor disagree	disagree	strongly disagree
	5	8	4	4	0

B) YOUR PERSONAL BACKGROUND

B1) What is your area of expertise and the role in your institution?
Please specify:
Area of expertise:
R1: River basin management
R2: modelling heavy metal emissions within river systems
R3: Nutriente balances in agriculture
R4: Water quallity and ecology
R5: Decision making, analysing waterpolicies
R6: Policy development/ Management/ Water
R8: river flood forecasting and modelling
R9:GIS and Remote Sensing
R10: Hydrology
R11: Forestry Sciences
R12: Consultancy Services in Environemental Management
R13: Fluid Mechanics; Hydraulics; Fluvial Hydraulics; Water Resources (Transboundary Rivers) and Environment
R14: Water resource planning, engineering and management
R16: Hydraulic Engineer/Reservoire Operations
R17: Water resources
R18: IRBM, environmental planning
R19: Water management, mainly in agricultural systems
R20: Modeller, hydrology, organisational psychology
R21*: Engineer
R22*: Engineer
R24*: Biologist
R25: Freshwater ecology, eutrophication
R26: Sustainable agriculture, science and community interfaces

Your role in the institution:
R1: Associate Professor
R2: Research Assisatant
R3: scientific expert in the field of agriculture and environment (nutrient balances, biodiversity and landscape impacts, organic farming

R4: Internal adviser
R5: Sr. Project manager Water management
R6: Project/Project leader
R8: Scientific Director
R9: Head of Cartographical Unit
R10: Staff scientist
R11: Director
R12: Director
R13: Professor
R14: Team leader
R15: Hydrology, water policy
R16: Reservoir Operation modeller
R17: Researcher
R18: lecturer, researcher
R19: Researcher (my Phd was the MERIT model for La Mancha Oriental Aquifer)
R20. Senior Advisor
R21*: Director of the technical/management sector
R22*: Member of the Administration Council
R23*: Director
R24*: Scientific Director
R25: environmental scientist
R26:Science Leader, AgResearch (New Zealand's largest Crown Research Institute, with research responsibilities across all fields of knowledge and value added for agriculture).

B2) Your involvement in the development/acquisition included:

Specification of the requirements, identification of the current and potential needs	20
Adaptation of the system to fit the data flow and systems/databases in your institution	12
Testing and evaluation of the prototypes at various development stages	16
Application of the system in practical situation before the system/acquisition was completed	11
Consultation with other potential users, or parties involved or affected by future application of the system	10
Training of technicians	7
Neither of them	3

Other, please specify:

B3) Since when have you been using the [DSS] system and how frequently?
R1: 2001, ongoing
R2: Since 1999, all the time
R3: 2004: once for the implementation of the WFD, 2005/2006: within the European wide project EUROHARP: MONERIS was applied by the German modellers to the Austrian river catchment "Gurk"
R4: We haven't used it until now
R5: Summer 2005, weekly
R6: still in progress
R10: Not since the project was stopped
R12: since 2003 on a continuous base
R13: Since 2003, on a a weekly basis
R14: Since approximately 1999 - we have developed four model applications (planning, forecast, accounting and daily) that are used occasionally to daily
R15: I was originally responsible for choosing and developing a new DSS in the early 1990's; throughout that decade a sure transition was made; I'm now the manager and this responsibilities are within my (?) Since 1996; used every day by my water operations department
R16: Since 1996-at least weekly
R17: 8.2005, 5hours/week
R18: regularly since 5 months
R20*: 2002, since a year
R21*: Since it has been acquired whenever necessary
R22*: Since 2001, 2 application
R25: Since 2004, several times
R26: We have seen the prototype kerDST system in use in the SRDTOOLS project in Europe during 2005, and saw its prospects for the stakeholder based regional futures studies and sustainable agriculture evaluation work that AgResearch is engaged in. We have by now incorporated use of the kerDST tools in several projects including "Choosing Regional Futures" (led by the Environment Waikato, the regional council concerned), and a set of industry sponsored projects intended to define methods for incorporating environmental and social performance considerations into assessment of agriculture strategies at farm, district and regional scales, e.g., river water quality and chemical run-off from intensive dairy farming.

	Yes	Not
B4) Are you a direct user of the system (i.e. you interact with the system installed on your PC)?..	17	9

	strongly agree	agree	neither agree nor disagree	disagree	strongly disagree
B5) It was difficult to get familiar with the system................................	0	8	7	6	2
B6) Your role in the process of decision making changed because of the adoption of [DSS]............................	1	4	5	9	4

If yes, how?

R5: Making more rational decision, better options for public participation, more insight in dc problems en solutions for the WFD (for example combined effect of several measurement)
R13: Indirectly, though the joint work with Regional water Institutions that began to use DSS for specific simulations
R14: The system is very complex and capable, like a CAD drawing program, but simple applications are possible if your needs are less complicated. RiverWare is a policy model, so it is most suited to systems requiring the implementation of complex operating criteria with many potentially conflicting conditions.
R15: Due to our ability to identify and compare alternatives, my role is primarely leading the environmental impact satement process to adopte new operational guidelines (in addition to managing the operation from Lake Moad (?)/ Hoover Dam to Mexico)
R19: We are offering solutions to current situations of non sustainable use of groundwater
R26: The kerDST system seeks, by design, to open up evaluation of options and also contributions to information sets and indicators, to a wide spectrum of users – usually stakeholders in the problem. This means that our role in AgResearch as experts is changed to become more one of facilitation than of execution of the method. Ideally, these tools help us to develop and then maintain involvement of members of the business and wider communities around the planning or business choice questions (such as taking environmental quality into account in farming). This means that we are involved in structuring a wider governance and monitoring process, not just specific decisions.

C) IMPLEMENTATION, USE AND MAINTENANCE

C1) What was the importance of the [DSS]' results for the policy decision to be taken?

[DSS]' results informed the decision (choices of the policy) but the choice has been taken independently.. **6**

[DSS]' results have been (more or less) directly taken up by the decision/ policy choice........	**9**
[DSS] was useful in a broader context, facilitating the planning process and learning among the actors, contributing to compromise different expectations and interest........	**14**

Other, please specify

R3: MONERIS was useful for a first status quo-survey on diffuse nutrient inputs into river systems
R4: 1. We expected that the WFD-Explorer will give insight in the relation of important measures and the ecological impact. This will help 'us'inmaking policy decisions; 2. The WFD-Explorer will fauhtate the decisions with other stakeholders.
R5: (note: answers are based on expectations!)
R9: It hasn't been used by users and Stackeholders
R14: We are using RiverWare for multiple decisions over multiple time scales – it was not a single-use DSS
R15: RW is also usefull in a specific context, being used each day for water scheduling, but also used to analyze specific impacts from different decisions fro our entire hierechy of decisions
R20*: It has also facilitated cross-cutting work with experts from other territories
R26: Note: kerDST is called a "Deliberation Support Tool" and it is meant to promote what we in New Zealand call collaborative learning. The regional planning legislation requires extensive community consultation for setting high level goals as well as for many individual project assessment or policy assessments. This means it is intended to help improve decision processes, not take them over.

	Yes	Not
C2) The system described above is still maintained and further developed within your institution or in collaboration with the developer	**20**	**4**
C3) Have you applied the [DSS] in more than one decision context?	**10**	**14**

If yes, how many times?

R5:20 different situations are expected
R8:2
R12: Three times
R14: We use it on a daily basis, as well as regularly and occasionally
R15: For 4 different EIS processes in the past 4 years
R16: two

R17: 2 times
R21: 3
R26: At present, we are using the kerDST tools in four distinct projects that deal with agriculture activities and choices at different spatial scales (farm scale, district or sub-catchment, and finally regional scale) and also contrasting timescales (long term planning or on-going monitoring). We are also considerijng the use of kerDST for another project on biosecurity in New Zealand.

	strongly agree	agree	neither agree nor disagree	disagree	strongly disagree
The re-use of the system was simple	6	5	6	1	0

Please, insert your comments:

R3: It was not simple, but after a three-days training on MONERIS we could apply it. Questions could be answered by e-mail and phone
R6: Sustem still in development for regional use
R9: It was useless
R10: We were not able to get this far
R14: It could be simple. Due to our extreme range of possible alternatives, operating criteria, and legal constraints, reuse difficulty varies with the model/application intent
R15: Our susytem's policies are quite complex; RW has been able to andle any new proposed policy for detailed analysis
R19: reuse easy for persons for high level of preparation. Anyway the model has been developed to hel water managers and big associations of stakhoòders which have adequate personal to use it. The model has not been developed for managing a farm
R26: Once you've used the Deliberation Matrix in one situation, it is very easy to see how to use it again

D) BENEFITS AND SATISFACTION

	strongly agree	agree	neither agree nor disagree	disagree	strongly disagree
D1) The [DSS]' implementation in policy making improved the usual/traditional approach of your Institution/Firm	9	11	3	1	0
D2) You are satisfied with the various components of the system (Leave blank if not applicable)					

Database management system..........................	5	8	1	1	0
Simulation model(s) (please specify below if different models are used)...............................	7	8	0	1	0
User interface...	6	10	2	4	0
Module facilitating the evaluation/ranking of policies..	7	5	3	3	0
Module responsible for documentation and presentation of the results...............................	4	9	5	1	0
Interactive exploration of the model suitability..	5	8	4	2	0
Role attributed to various interests groups (stakeholders)..	5	7	2	2	0

What were the most effective components and for what reason?

R1: retention approach in groundwater and surface water because it is essential for the results and is highly innovative
R2: the operational approach for calculating the input of substances to surface waters; the organisation or the model on the level of subcatchments; the aggregation along the flow path of the river systems.
R3: Nutrient balance component on political districts, because this was renewed with Austrian nutrient coefficients and first time done on regional level (political districts)
R4: 1. the knowledge database with dosis-effect relations for translation of measures to effects on ecology; 2. a fast way to calculate changes in water quality
R5: Coupling detailed models and databases in simplified way, so the results (of several policy alternatives) are calculated instandly
R6: Visual presentation of policy decision also meant to communicate with stakeholders
R8: Module facilitating the evaluation of indicators (value function) for the involvement of stakeholder options in a participative way
R9: No Opinion
R12: The capability to perform a number of assessments of the water system of the study area in terms of sources, usage, environmental quality, evaluating the effects of actions on the basis of different scenarios, alternatives and policies
R13: The most effective components for the work undertaken were the allocation and the economic modules as they were adequate to the needs of our case study. The allocation module allowed the user to analyse on a monthly basis different aspects of the water allocation as the water demand and water deficit. An important aspect is the fact that the user can obtain the characteristics of the water conveyance and supply at each step of the water network. Moreover, the economic module performs a simple economic analysis,

which is very helpful for options and strategies comparison. Moreover, the calculation of the rate of cost recovery is beneficial to cope with the WFD requirements.
R14: The ability to accurately capture policy and legal constraints in rule statements
R15: the ability to enter new policies and the comparison of those policies has allowed us to find innovative solutions to complex problems
R18: Simulation of measures and its interaction in different system fields, visulisation with maps and graphs
R19: Module to compare options in order to assure sustainability
R21*: The database allows for a rapid synthesis and elaboration of a huge amount of data; the module for evaluation and classification of the policy options allows for an interdisciplinary approach to a complex problem
R22*: Definition and role of Stakeholders
R25: User interface and module facilitating the evaluation/ranking of policies, because of their user-friendliness
R26: The main focus of kerDST is to build dialogues between stakeholders and with scientists and technical staff of the institutions involved and to make comparative assessment of policy options. We know that the Deliberation Matrix can help a lot with these two tasks, e.g., highlight for decisionmakers and stakeholders what are the trade-offs in farm business or regional planning decisions. The kerDST system can also help with aspects of data base management, documentation and model suitability assessment, partly through being embedded in interactive multimedia on line. However these functions are more intensive and we have not yet progressed far enough in the projects to give in-depth feedback.

and what were the main limitations and for what reason?
R1: Erosion and particulate transport of P in river systems- the knowledge is generally rather limited an these processes
R2: limitations for our work belong mainly to input data, which is not available and data uncertainties and not to the modelling approach
R3: Soil map, soil erosion (P): no detailed data. Drainage areas: no data. Snow melting processes: not modelled within MONERIS, estimation on surface runoff.
R4: The knowledge database was not complete for all artifical and heavily modified waterbodies (types/kinds)
R5: Although it is simplified it takes time for policy makers to work with the system!
R6: Scale (winter versus summer), no more detailes information
R7:
R8: including the indicators in DPSIR chain due to some ambiguities related to including an indicator in P or S
R9: Stackeholders and uswes opinions have not been taken into account
R12: Data was not always available

R13: The main limitations are related with the the fact that it is not an open source software.
R14: Technically complex with a significant learning curve for complicated implementations
R15: no (?) limitationa
R18: Evaluation of measures is too complicate
R19: The models runs just for 1 year no more.
R20: Structural learning
R21*: Concerning the users interface, it is necessary to improve the videoconferencing part
R22*: The system is not that practical for use. (No friendly)
R24*: The interactive exploration and classification of the various options: these allow to point out the various roles and the background knowledge of the various Stakeholders
R25: Nothing particularly
R26: Like all tools, people have to be motivated to take the time to learn their use. Some of our partners are very busy in industry, farming, administration or community life, and so they do not invest as much as would be ideal. This is not really a limitation of the tools but it is a real limit on effective use.

Additional comments:
R14: If your system and / or needs are complicated and policy driven and /or legally constrained, this DSS will likely satisfy you
R18: systems integrates really the most important physical and economic models/ issues if IRBM
R19:The importance of transparence and PP to allow the effectiveness of the application and of the results
R20: Takes time to get familiar with probabilities and what BNs mean
R21*: Obstacles and interferences from Stakeholders not relevant to the decision to be taken
R24*: The non-applicability of the results at a political level: even if transparency in the method was assured; its authoritativeness wasn't not recognized by everybody
R25: For our institution, CatchMODS integrated with a reservoir-plankton-model was necessary to demonstrate an environmentally friendly way of managing a receiving-water environment as part of pollution reduction program so the whole modelling framework itself was very important to show councils and stakeholders to find alternative decision choices and their likely benefits in concrete terms (e.g. dollar values), all of which CatchMODS could deliver.
R26: People often want community consultation and public participation to be done fast and at low cost. They have the idea that these things should take just a

few days. Although the Deliberation Matrix can be used to make a surface assessment of options in just a few hours, full use of the full system can engage a team for many weeks. Then we run up against the perception that everything should happen straight away, which is not expected for technical research and modelling, for example.

	strongly agree	agree	neither agree nor disagree	disagree	strongly disagree
D3) The use of the system or the quality of its components were contested by those who were affected or involved in the decisions made with help of the DSS	2	8	5	3	4

If yes, what were the main contestations?
R1: Quality of input data.
R4: 1. Some people in the institution think that the user interface is to full and complicated for stakeholders; 2. the rest we don't know yet.
R5: Clear and simple user interface, presenting the results in a way it can be understood by people not familiar with the context (e.g. WFD).
R9: The Portuguese partners didn't took into account our points of view
R12: For the DSS end user (government policy makers) the system was a BLACK BOX, so they did not trust the results and conclusions immediately
R13: Some inadequacy of the DSS tool requirements to the available data (mostly for the quality module).
R15: We formed a stakeholder group and informed them throughout our process of (?) to RW
R20: Gave a full picture and flexible tool.
R21: Reducing the number of actors involved in the tools' development phase. Improve the testing during the development phase.
R24: The role of the various stakeholders wasn't recognized.
R26: Not applicable – everyone with whom we have discussed has reacted positively to the tools and their goals for helping to open up evaluation processes.

D4) Do you have any recommendation for the improvement of the development of similar systems in the future?
R1: Improvement in the software implementation, Implementation of GIS interface.
R2: Improvements should mainly be done on the European wide and consistent supply of input data
R3: more user-friendly surface to improve handling of model inputs and outputs
R4: 1. Complete the knowledge database of doses-effect relations for al kinds of artificial and heavily modified water bodies; 2. Make the user interface more

clearer for communication with stakeholders; 3. Connect the knowledge database of dosis-effect relations directly to water quality models (for example: SOBEK)
R5: See D3. Developing the DSS together with the end users
R9: Technical follow-up of the Project, has to be carried in a more careful way
R10: I hope to able to use the mDSS in a future project
R13: The use of the WSM DSS tool for the WSM Project purposes and further investigation allowed us to understand the difficulty of such a model to suit everyone's needs. In fact, it seems that each decision support system tool should, ideally, be adaptable to different scales of analysis and databases. Open access to source codes should help.
R14: RiverWare is continuously under development through feedback and requests from users for additional features. The software was designed to be extended and has successfully been enhanced – integration with other models is now underway. Regular training on new features, if useful, should be planned
R15: This technology is expensive to develop and to maintain. Partnership between larger Institutions will be critical to fund the next generation of RW-type tools. The "model used to develop RW (collaboration/partnership between Univ. of Colorado Bureau of Reclamation and Tennessee Valley Authority) worked well, including licensing the product to others to help offset maintenance costs.
R18: better interface, more flexible data base interface
R19: As a dev he did a great job
R20: There could be a data base with all used data and assumptions so that it is easier to update the BNs
R22*: no
R24*: The method is a powerful tool. It could be eventually improved in its interface to allow a better discussion on the alternative options
R25: Good communication is a key to integrate our needs in the system to be developed. I believe we maintained good communication throughout the course of the project, resulting in successful development of CatchMODS and its presentation to policy makers
R26:We chose this DST system because it fills a gap between production of technical knowledge and building dialogues with stakeholders. For example, in the Waikato "Choosing Regional Furtures" project, we have partners with good system modelling and spatial analysis skills, but we needed to have a way of interfacing the modelling with the stakeholders' perceptions of issues for the region's future. The Deliberation Matrix gives us a solution for this. In the case of work with farmers about integrating social and environmental considerations into assessment of farm management systems, the Deliberation Matrix framework helps to present the diversity of stakeholders and their differing concerns in a non confrontational way. The kerDST system is quite new and some of its on-line

interfaces can still be improved. But overall we are satisfied with its potential and we expect to keep using it for the next few years.

NOTE: (*) the answers given by the respondent (R) have been translated form Italian into English

IWA Publishing's authorised EU representative for General Product Safety Regulations is Diane D'Arras, 15 rue Duret, 75116 Paris, France, e-mail: safety@iwap.co.uk.

Printed and bound by CPI Group (UK) Ltd, Croydon, CR0 4YY
12/05/2026
02108891-0002